Praise for
Planting an Idea

"*Planting an Idea* carries a critical message about global sustainability and environmental justice while suggesting a practical and hopeful approach to solving the world's most pressing issues. Jerry Apps's historical storytelling combines with Natasha Kassulke's fine journalistic writing to entertain while they teach. Most importantly, each chapter shares lessons to help bring people together, a welcome antidote to our divided nation's public discourse. A must read for book clubs, students, and community leaders!"

**—Terry Daulton, environmental educator, artist,
and founding president of Wisconsin's Green Fire**

"There is no time like the present, and with ever-increasing levels of population growth, consumption, and technology, the present is critical. So just when we needed it most, Apps and Kassulke have given us *Planting an Idea: Critical and Creative Thinking About Environmental Problems*. Decidedly not esoteric, we now have a guide to understanding the environmental issues of our time, but we also have a pragmatic text that shows how to evaluate and act on these threats. *Planting an Idea* shows us how to live on our home planet and have a future. We all need to read, and apply, this book to our lives!"

**—Richard Knight, Professor Emeritus
of Wildlife Conservation, Colorado State University**

"What happens when one of Wisconsin's great storytellers collaborates with a top-notch environmental journalist, who just happens to be his daughter-in-law? The resulting book gives the reader a fresh take on environmentalism, with an emphasis on hopeful stories from Wisconsin. It's neither a history textbook nor simply a celebration of Wisconsin's conservation traditions. Rather it engages the reader and helps point a way forward in times when environmental threats can seem overwhelming. Wisconsin's great conservationist and writer, Aldo Leopold, took a somewhat similar approach in his classic essay, "Good Oak," and now *Planting an Idea* adds to Wisconsin's rich tradition of environmental literature."

—Stanley A. Temple, Beers-Bascom Professor Emeritus in Conservation, University of Wisconsin–Madison, and Senior Fellow, Aldo Leopold Foundation

PLANTING
AN IDEA

PLANTING AN IDEA

*Critical and Creative Thinking
About Environmental Issues*

**By
Jerry Apps and Natasha Kassulke**

Fulcrum Publishing
Wheat Ridge, Colorado

Library of Congress Cataloging-in-Publication Data

Names: Apps, Jerold W., 1934- author. | Kassulke, Natasha, author.
Title: Planting an idea : a guidebook to critical and creative thinking
    about environmental problems / by Jerry Apps and Natasha Kassulke.
Description: Wheat Ridge, Colorado : Fulcrum Publishing, 2023. | Includes
    bibliographical references and index.
Identifiers: LCCN 2022041308 (print) | LCCN 2022041309 (ebook) | ISBN
    9781682753422 (paperback) | ISBN 9781682753644 (ebook)
Subjects: LCSH: Environmentalism. | Environmental protection. |
    Environmental degradation. | BISAC: SCIENCE / Global Warming & Climate
    Change | BUSINESS & ECONOMICS / Environmental Economics
Classification: LCC GE195 .A66 2023 (print) | LCC GE195 (ebook) | DDC
    363.7--dc23/eng20230111
LC record available at https://lccn.loc.gov/2022041308
LC ebook record available at https://lccn.loc.gov/2022041309

Printed in the United States
0 9 8 7 6 5 4 3 2 1

Cover art and design by Kateri Kramer
Interior illustrations by Kateri Kramer

Unless otherwise noted, all websites cited in endnotes were current
as of the initial edition of this book.

Fulcrum Publishing
3970 Youngfield Street
Wheat Ridge, Colorado 80033
(800) 992-2908 • (303) 277-1623
www.fulcrumbooks.com

Contents

Introduction

Just Think About It

Why write this book? And why now? For us, as environmentalists, the answers to these questions are black and white. From our perspective, we are at a critical time in the history of our planet. We are at a tipping point that is leaning toward disaster. But we also believe that we can stand up for the environment, and we can straighten out this calamity.

To some, this response might sound a bit—perhaps a lot—optimistic. But it is vital. In the planet's long history, concern for the natural environment has never been more important. Today, a strong argument can be made that no corner of the globe is immune from environmental degradation, as evidenced by widespread climate change impacts. Across the world, we face water quality and quantity concerns, land use conflict, air quality degradation, rampant species loss, and food insecurity.

As a result, we are calling for bold collective action based on facts that inform critical and creative thinking about solutions.

But we greenies cannot do this alone. We need people from a variety of social and ethnic backgrounds; of different genders, ages, educational backgrounds and affiliations; and diversity in all areas of human nature. We are at a time when policy makers, politicians, environmental organizations, and individuals must work ever harder and together to protect the future for our grandchildren and the generations who follow.

Yet, never in our lifetimes have we seen such dramatic divisions between those who want to protect the environment and those who wish to exploit it for its monetary return. In the 1960s, when environmental protection was a topic for widespread discussion and action, both conservatives and liberals worked together to fashion legislation to improve and protect water, air, land, and wild creatures.

Since the Great Recession (2007–2009), policy makers have taken sides—often extreme sides. On one side are certain conservatives who view the environment as an impediment to economic development and job creation. On the other side are certain liberals who wish to protect the environment at all costs. The two groups are at loggerheads, with considerable emotion and strong opinions shared and debated, often with the facts of the matter ignored or manipulated to fit a particular point of view.

What is needed, in our judgment, is for both groups to stand back and examine their basic beliefs and values about the environment and do some critical and creative thinking, followed by action. We also need long-term planning that ensures environmental and economic policies are centered around health needs—both ours and the planet's.

While we come from different genders, educational backgrounds (Natasha has studied mass communications and biology, and Jerry holds a PhD in education and rural sociology), life experiences, and generations, we agree that thinking without action accomplishes little. And action without careful thinking is dangerous and often futile.

Aside from the various groups that have taken sides for and against the environment, each one of us, every last person, has an individual responsibility for caring for the environment. Belonging to an environmental group is a good idea. But concern for the environment goes beyond concerned groups. It is everyone's responsibility, whether working alone or working with others.

We don't mean taking some willy-nilly action without first examining the position you are taking. It does not mean joining a group that appears to agree with your position when you have not carefully examined what your position is.

This book is designed to help you figure out what your position is on a particular environmental problem, and ultimately not only know that position, but also help provide evidence to back up your position. And not just any evidence, but accurate, verifiable evidence from reputable, reliable sources. So, in a way this is a guidebook for examining, and thinking critically and creatively about the important environmental problems that face our planet today.

We have both contributed to each chapter of this book, but have divided up the primary writing responsibilities for those chapters based on our experiences and expertise. Here is the lineup of who was the primary author for which chapter.

For chapter 1 ("The Environmental Movement: Providing a Brief History"), Jerry has set the stage by writing a brief history of the environmental movement and how we've gotten to where we are today regarding the country's concern, or lack of concern, for environmental problems. For chapters 2 ("All About Thinking: Evaluating How You Know What You Know") and 3 ("A Game Changer: Combining Critical and Creative Thinking"), Jerry delves into the meaning of, and how to do critical and creative thinking. We believe that by carefully examining environmental problems through the lenses of these two types of thinking, progress can be made in helping solve many of the issues facing our natural environment today. Jerry is also the lead author for chapter 5 ("Agriculture: Growing from a Complex History, Facing an Uncertain Future"), as well as chapters 6 ("Forests: Discovering Deep Roots and Branching Out"),

7 ("Water: Riding a Wave of Complex Issues") and 8 ("Energy: Building Up Steam for Sustainability").

Natasha wrote chapters 4 ("Climate Change: Finding Our House Is on Fire"), 9 ("Air Quality: Finding It's Too Soon to Breathe a Sigh of Relief"), 10 ("Natural Resource Issues: Where Enjoying Meets Exploiting"), 11 ("Land Use: Seeing the Need for Resilient Road Mapping"), 12 ("Endangered Species: Going, Going, Gone"), and 13 ("Biodiversity: Protecting Healthy Ecosystems").

We share the writing of this introduction and the concluding chapter ("Just the Beginning"). There is an extensive references section, with books that describe in great detail how to do both critical and creative thinking, as well as books that tackle the various challenges currently facing the natural environment.

Jerry Apps
Natasha Kassulke

CHAPTER 1

The Environmental Movement

Providing a Brief History

In the United States, the identification of an environmental movement is generally traced to the 1960s. But concern for the environment goes back many years before that. Winona LaDuke, a Native American, writes about the concern Native people had for the environment from their earliest days on the land. The Native Americans' concern for the environment goes back to the Indigenous philosophy of a "reverence for the interconnectedness of all life forms."[1]

The coming of the Industrial Revolution (1760–1840) resulted in considerable environmental pollution, especially air and water pollution. Several groups were formed in response to these environmental challenges. One was the Emergency Conservation Committee, founded by a New York socialite, Rosalie Edge, in 1929.

The Great Depression (1929–1941) not only resulted in an economic calamity with millions of people out of work, it also revealed what would happen if grassland not especially well-suited to agricultural crops was plowed and planted, and then the needed rains did not come. Starting in about 1930 and continuing through much of the Depression years, these once stable grasslands, many in the southwestern states, now cultivated for crops, saw massive dust clouds, created by the dry winds from the West that arrived and refused to leave. The cry for soil conservation strategies began emerging. Then President Franklin Roosevelt signed the Soil Conservation Act on April 27, 1935, which created the US Department of Agriculture's Soil Conservation Service.

Two years earlier, in 1933, President Roosevelt and the Congress created the Civilian Conservation Corps (CCC). Unemployed young men, organized into camps of about 200 each, planted trees, built dams to prevent soil erosion, advocated contour farming, helped construct state and national parks, fought forest fires, and participated in many other environmentally focused programs. The CCC, which operated from 1933 to 1942, helped build an awareness of the environmental challenges the country faced and set out to correct them.

Following the end of World War II, dramatic changes began occurring in this country—changes that in many ways went well beyond the obvious advances in science and technology. For example, electricity became widely available, and agriculture changed from small family farmers to larger holdings focused on single enterprises such as dairy, poultry, hogs, or grain crops. In contrast, the small family farms were diversified with several small enterprises.

As a result of these revolutionary changes in agriculture, thousands of rural people fled to the cities in search of jobs and a new life away from the land. When they accumulated enough money for a down payment, they left the inner cities and built homes in the surrounding areas, on farmland that soon became extremely valuable as homesites.

As the rural population dwindled to a fraction of what it was in the 1930s and 1940s, and the resultant urban population increased, more and more people had little or no firsthand experience with the natural world. What they knew about nature they heard from the stories their parents and grandparents told of life on the farm, or from a book they might read, or a TV show or movie they might see. Nature had a different meaning for most urban people who had not experienced it directly. Many were interested in learning more about it to be sure—bird-watching clubs emerged, and communities developed parks and nature preserves where people could hike and observe nature. Author David Suzuki wrote, "The most destructive aspect of cities is the profound schism created between human beings and nature. In a human-made environment, surrounded by animals and plants of our choice, we feel ourselves to have escaped the limits of nature."[2]

By the early 1960s, hordes of people drove gas-guzzling automobiles that were as large as hay wagons and spewed exhaust pollution into the atmosphere with no concern for the future. Technology was on a roll—new insecticides, such as DDT, emerged—a savior spray, some said. A way to control disease-carrying mosquitoes and pesky flies. Unbeknown at the time, DDT also prevented millions of raptor eggs, such as those of the bald eagle, from hatching.

Water and air pollution—the results of the country's so-called progress—had increased to dangerous levels. Blatant use of pesticides ravaged certain populations of wildlife. Land conservation efforts were just beginning in many parts of the country—with still-fresh memories of those Dust Bowl years of the 1930s when millions of acres of topsoil blew away on the westerly winds that raged across the central and southwestern parts of the country.

One time during the early 1960s, I was driving around Gary, Indiana, on an early evening and the sky was a sickly red from the manufacturing plants operating there at the time. Also, around that time, I was driving in West Virginia and came upon entire hillsides of dead trees—killed by the effects of acid rain, I was told. The acid rain resulted from coal-burning power plants located many miles to the west spewing sulfur-laden coal smoke into the air.

People with breathing problems suffered, animals suffered. Fish from polluted rivers and lakes could not be consumed. Birds were dying. But most people were slow in realizing that unless something changed, and quickly, nature would evoke its wrath on the people—as it was already beginning to do.

Some people, mostly students in high schools and colleges read Henry David Thoreau's *Walden* but found it difficult to understand, or simply dismissed his words as those coming from someone living at another time, in a different place, who wrote with long sentences and described a quaint lifestyle that was no longer found in our modern world.

The idea of being concerned about nature was not on anyone's mind. People were too busy buying things, trying to amass wealth, raising families, purchasing homes—and wondering, and sometimes worrying about and participating in the civil rights movement and the anti–Vietnam War protests that had begun to erupt on college campuses in the 1960s across the country.

By the early 1960s, "planned obsolescence," a term used to describe products that were built to wear out so they could be replaced and thus keep people employed in making replacements, became popular. Everything from cars to radios, kitchen appliances to ballpoint pens were manufactured with a limited lifetime.

Advertising began shouting to a population that it must not only buy a certain product, but that it should replace an older model with a newer, more modern one. "Keeping up with the Joneses" was a phrase tossed around in those days, which meant if your neighbor had purchased something, say a new car, you should do so as well, and preferably, your new car should be a larger, more expensive one.

When my father bought his first farm tractor in 1945, a 1943 model Farmall H manufactured by the International Harvester Corporation of

Chicago, no one hinted that the tractor had a certain life expectancy and that one day my dad should look to replacing it. My dad expected the tractor to last, and last it did. When he sold the home farm in 1973, the Farmall H was still quite capable of doing everything it could do when it was new. It was clearly built to last.

By the 1960s, a change in the mindset of people was well in place, an emphasis on consumerism began—some of it a response to the World War II years of scarcity and rationing when people had to make do with little. The economy depended on people buying things, whether they were needed or not.

The 1960s and early 1970s saw the country tearing itself apart. Despite President Lyndon Johnson signing the Civil Rights Act of 1964, which was a major step forward, the tumult of the civil rights movement continued, there were concerns that communism would take over the world, thousands of college students were protesting, and there were riots in the major cities. With all of this unrest, mounting problems with the environment were overlooked.

It wasn't until the late 1960s that public concern for the environment became evident in this country. A renewed awareness for nature and the environment had emerged when Rachel Carson's *Silent Spring* became a *New York Times* best-selling book in 1962. The book sold more than a half million copies in twenty-four countries—and for the first time the word "environment" was used in more than just spelling bees, as one writer pointed out.[3]

Rachel Carson wrote how chemical pesticides, especially DDT, although good at killing flies and mosquitoes, had horrendous side effects that were resulting in the demise of bald eagles and other raptors, as well as causing cancer. As the bald eagle is the national bird of the United States, its potential downfall led to considerable public attention and concern for the preservation of wild spaces and such laws as the Endangered Species Act (1973).

In addition, by the 1960s, the US public was becoming concerned with the increase of pollutants in the environment: factory-polluted rivers and streams, auto emissions, and oil spills as examples. Also by the 1960s, the pollution of the Great Lakes was drawing public attention, especially Lake Erie and the Cuyahoga River that drained into Lake Erie. On June 22, 1969, a spark from a train traveling on a bridge over the river fell on the industrial waste floating in the river and created a fire that roared to five stories tall.[4] On January 28, 1969, an oil well off the coast of Santa Barbara, California, suffered a blowout, resulting in an oil spill that sent 3 million gallons of crude oil into the Pacific Ocean. It was, at the time, the largest oil spill in US waters. People saw on their televisions and newspapers photos of oil-covered birds and dead dolphins. They became outraged, as many worked to clean up the beaches. People also began lobbying oil companies to stop this horrific pollution.[5]

Wisconsin's then-senator and former governor Gaylord Nelson was listening to what Rachel Carson and her advocates were saying. This humble man from Clear Lake, Wisconsin, suggested grave consequences for the future of this country if something was not done, and done quickly about the mounting environmental challenges. In response, Senator Nelson organized Earth Day, which was held on April 22, 1970. Some 20 million Americans "took to the streets, parks, and auditoriums to demonstrate for a healthy, sustainable environment in massive coast-to-coast rallies."[6]

On that April day in 1970, I sat in a jam-packed Stock Pavilion on the University of Wisconsin–Madison campus and heard Senator Nelson proclaim the necessity for everyone to not only become more aware of the environmental problems facing the nation, but to do something about solving them. To push our lawmakers toward passing legislation to correct the errors of the past and assure they would not occur in the future. In a rare moment of solidarity, Republicans and Democrats, rich and poor, young and old became convinced that nature could no longer be ignored, nor the environment taken for granted. Thousands agreed, including many lawmakers, that major action must be taken to stop air and water pollution, save endangered species from extinction, and prevent indiscriminate use of pesticides and a host of other concerns.

A series of actions occurred. In 1970, President Richard Nixon established the US Environmental Protection Agency (EPA). That same year, not long after passing it, Congress amended the Clean Air Act to create national standards for air quality, auto emissions, and anti-pollution. In the succeeding years a series of other laws were passed—1971, banning of DDT; 1972, agreement between the United States and Canada to clean up the Great Lakes; 1972, the Clean Water Act; and more.

The 1970s saw broad bipartisan support for environmental goals. This changed with the election of Ronald Reagan in 1980. President Reagan had a probusiness approach, which often challenged regulating businesses to improve environmental conditions. The environmental movement suffered. But the US population largely continued supporting environmental goals.

By the 1990s, the environmental movement began focusing on broader, international issues including global warming, acid rain, biodiversity, ozone depletion, and destruction of rain forests.[7]

By the beginning of the 2000s, and especially after wars in Iraq and Afghanistan sucked up billions of dollars of resources and too many lives, a devastating recession began in 2007. More people turned away from nature and concerns about the environment. Jobs and economic development became the focus—caring about the environment moved from becoming a necessity for the future of the planet to an inconvenience that, in the minds of some, too often prevented job creation.

By the mid-2000s, the environmental movement had become highly politicized. Taking care of the environment became a "bogeyman" for an expanding conservative movement in the country. Loud-talking radio commentators, conservative bloggers, and a host of others derided the environmental movement as one associated with liberals and progressives "who were more concerned about saving an endangered butterfly than creating jobs."

Thus, the environmental movement stalled—pushed aside by those who believed short-term goals, such as job creation, efficient use of environmental resources, and helping the nation become energy independent, were far more important than worrying about the long-term effects of global warming, water shortages, air pollution, hunger, poverty, and a host of other challenges facing the planet.

Then, in late 2019, the COVID-19 pandemic took first place in the minds of many—and environmental concerns stayed in the background.

These days, people often take the natural environment for granted, find it an annoyance, and some believe nature blocks their quest for riches and related human endeavors. Unfortunately, many people have ceased to realize how important the understanding of, and appreciation for, nature is to the future of us all. People may see nature as a park to visit, a river to canoe, a bird to watch, or a nature preserve trail to hike. These are laudable activities and are not to be downplayed, but if the natural environment is to remain as we have known it in the past, if our air and water are to remain clean, our land rich and prosperous, and thus our food abundant and safe, we must view nature as much more than merely a place to visit and study as a pastime, as a recreational activity. A healthy natural environment in the future requires much more of us than passive enjoyment.

When I published my first book in 1970, *The Land Still Lives* (Wisconsin House), I asked Senator Nelson to write an introduction to the book. Senator Nelson wrote:

> Today the crisis of the environment is the biggest challenge facing mankind. To meet it will call for reshaping our values, to put quality on a par with quantity as a goal of American life. It will require sweeping changes in our institutions, national standards for the

goods we produce, a humanizing of our technology, and close attention to the problem of our expanding population.

Most of all, it will require that the people assert their right to a decent environment and that they evolve an ecological ethic of understanding and respect for the bonds between man and his planet.

What Senator Nelson said in 1970 is as important today—perhaps even more important—than it was then. The Wisconsin Historical Society Press published a Fiftieth Anniversary edition of *The Land Still Lives* in 2019.

Many place the modern start date for the environmental movement as 1970, with the first celebration of Earth Day.[8]

As we examine and seek solutions to today's many environmental challenges, a combination of critical thinking and creative thinking is essential.

Environmentalists Paving the Way

Many men and women have helped and continue to help pave the way for a national environmental movement.

Aldo Leopold (1887–1948) was a conservationist, environmentalist, forester, philosopher, educator, naturalist, and writer. He is probably best known for his book *A Sand County Almanac*, which was published in 1949. He is considered the father of wildlife ecology, and taught the idea of a "land ethic," "which calls for a caring relationship between people and nature." Leopold is seen by many as the most influential conservation thinker of the twentieth century.[9]

Frederick Law Olmsted (1822–1903) is often referred to as the "father of American landscape architecture." He developed an interest in nature as a young child, and he is best known for his role in designing major urban parks, such as Central Park in New York City. Olmsted created city parks all across the United States and believed that urban people should have access to open spaces (parks) in their cities.[10]

Henry David Thoreau (1817–1862) had an enduring love for nature. His book *Walden* is still widely read. One of his famous quotations, "In wildness is the preservation of the world," is still held high by environmentalists.[11]

Jens Jensen (1860–1951) was born in Denmark, moved to the United States in 1884, and settled in Chicago. There he worked for the West Side Park System from 1890 to 1900. The Cook County Forest Preserve, one of the most extensive nature parks in the country, was largely his creation. In 1935, at the age of seventy-five, he established a Danish–style folk school, called The Clearing, in Door County, Wisconsin. "Jensen believed that environments have a profound effect on people and that an understanding of one's own regional ecology and culture is fundamental to all 'clear thinking.'"[12]

John Muir (1838–1914) grew up in Wisconsin and is best known as the father of the National Park System. He was instrumental in creating Yosemite National Park, the nation's third oldest national park, in 1890. He also helped create Sequoia, Mount Rainer, Petrified Forest, and Grand Canyon National Parks. Muir and some of his supporters went on to found the Sierra Club in 1892. One of Muir's biographers wrote, "He taught the people of his time and ours the importance of experiencing and protecting our natural world." Muir was also a prolific author.[13]

Rosalie Edge (1877–1962) from New York organized the Emergency Conservation Committee in 1929. Her group focused on expanding protections for birds with the goal of saving additional species.[14]

The first woman director of the US Fish and Wildlife Service, **Mollie Beattie (1947–1996)** was also an advocate for the National Wildlife Refuge System.[15]

Rachel Carson (1907–1964) was a major figure in alerting the public to the dangers of DDT with her book *Silent Spring,* published in 1962.

Lady Bird Johnson (1912–2007) President Lyndon Johnson's wife, promoted wide-ranging environmental legislation, including the Highway Beautification Act of 1965 and the Wilderness Act of 1964.[16]

Winona LaDuke, a member of the Anishinaabe tribe, is a Native American environmental activist. LaDuke advocates for environmental justice, concern for climate change, and many other current environmental problems. She is cofounder of Honor the Earth, a Native American–headed organization raising awareness for Native environmental issues; she is also a supporter of the Green New Deal. Inducted into the National Women's Hall of Fame in 2007, LaDuke is an inspiration to all women environmental activists.[17]

Carol Martha Browner was the second woman administrator of the EPA, heading the agency from 1993 to 2001. From 2009 to 2011, she was the director of the White House's Office of Energy and Climate Change Policy.[18]

Christine Todd Whitman was administrator of the EPA from 2001 to 2003. She was an advocate for the Clear Skies Initiative to protect clean air.[19]

An authority on atmospheric sciences and climate research, **Warren Washington**'s work focuses on computer modeling. The 2007 Intergovernmental Panel on Climate Change assessment used Washington's models for studying the impacts of climate change in the twenty-first century. For this work, Washington and fellow National Center for Atmospheric Resources scientists, along with other colleagues, shared the 2007 Nobel Peace Prize.[20]

Appointed by Barack Obama in 2009, **Lisa Jackson** was the first African American to serve as the federal EPA administrator. As administrator, she focused on those especially vulnerable to environmental and health threats.[21]

Hilda Lucia Solis promotes environmental justice within communities. She was the twenty-fifth US secretary of labor from 2009 to 2013 and served in the US House of Representatives from 2001 to 2009. She was the first Hispanic woman to serve in the Cabinet and the California State Senate.[22]

Dr. Robert Bullard, often referred to as the "father of the environmental justice movement," was named one of *Newsweek*'s "13 Environmental Leaders of the Century." His books address a wide variety of issues, including "urban land use, industrial facility siting, housing, transportation, climate justice, emergency response, smart growth, and equity."[23]

Al Gore, vice president of the United States from 1993 to 2001, wrote *Earth in the Balance: Ecology and the Human Spirit* (1992). In his 2006 documentary, *An Inconvenient Truth*, he explored global warming, and the film won an Academy Award for best documentary. He was awarded the Nobel Peace Prize in 2007 for his climate change activism.[24]

Carl Anthony founded the Urban Habitat Program, "one of the country's original environmental justice organizations and is the former head of Earth Island Institute." He's also responsible for spearheading the nation's only environmental justice periodical: *Race, Poverty and the Environment Journal*.[25]

Deb Haaland, a member of the Laguna Pueblo tribe, is the country's fifty-fourth secretary of the interior and is the first Native American Cabinet secretary. Her interests include environmental justice, climate change, and Native land rights. Haaland says: "We are in a new era now, and we must do all we can to live up to [the agency's] mission of managing and conserving America's public lands, natural resources and cultural heritage; and honor the trust and treaty obligations to the nation's 574 federally recognized Indian tribes."[26]

Greta Thunberg is a Swedish environmental activist focused on climate change. In 2018 she started Fridays for Future (also called School Strike for Climate). Her address to the United Nation inspired a global climate strike on September 20, 2019. She was named *Time* magazine's Person of the Year in 2019 and is credited with coining the term "climate strike."[27]

Each of these men and women had or has a following of supporters, and each has made a major contribution by offering a perspective on the importance of the natural environment and its contribution to humankind, and the need for people to care for the environment as it has cared for us.

CHAPTER 2

All About Thinking

Evaluating How You Know What You Know

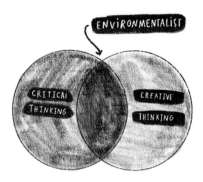

All of us do it, almost all of our waking hours. We can't help it; thinking is part of what makes us human. We think about tomorrow and what we should do and what we shouldn't do. We think about ten years from now and what we think we should be doing by then. We think about people who are close to us, our family and our friends. We think about yesterday, the joys and the sorrows. We think about what we should be doing right now. Sometimes we wish we could turn it off, flip a switch, and the thinking would stop.

So what is thinking? Here are two dictionary definitions:

"To employ one's mind rationally and objectively in evaluating or dealing with a given situation."[1]

"The *activity* of *using your* mind to *consider* something."[2]

Here are some specific applications. People ask us what we think about someone or something that has happened. They want our opinion. A friend asks us what we thought about a book we just read—once again an opinion is wanted.

When we need to buy something—perhaps a different car or a new piece of clothing—we must decide among many choices; we must sort out which of the sales pitches we hear and read are of help in making a decision and which are not. Which are truthful and which are not. We may ask a friend who has a make of car we are considering buying for her opinion. We are forced to think, to weigh facts and opinions, to decide which are accurate, and which may be inflated or downright false. Now we are doing thinking that takes place at a deeper level.

Problem-solving is thinking. When I was twelve years old, I got my first bicycle. I had saved money from picking cucumbers and green beans on the home farm to buy a speedometer for it—a device that would tell me how fast I was going and how many miles I had traveled. With my father, I stopped at the hardware store in Wautoma, Wisconsin, that sold bicycle accessories, including speedometers. The price was $4.95, as I recall. Arriving home, I was excited to attach my new speedometer to my bike. But I soon discovered a part was missing, making it difficult to fasten the speedometer to the bike's handlebars. I mentioned this to my father, who said that he did not have time to return the speedometer to the store in Wautoma, some twelve miles away. So, I was faced with figuring out what to do. I could wait until Dad would make another trip to Wautoma, which might be two weeks or more. Or I could figure out a way to fasten the speedometer to the handlebars without using the piece designed to do that. I went to bed that night, disappointed and disgusted, having spent some of my hard-earned money on something that I couldn't use. As I was going to sleep, I came up with an idea that I thought might work. The next morning, I tried the idea, using some of the parts in the speedometer package in a different way, I managed to figure out how to fasten the speedometer to the handlebars and make it work, which it did very well.

As a kid growing up on a farm, I faced many problems that needed solving, ranging from teaching a reluctant calf to lead so it could be taken to the fair, to how to split a block of wood that seemed to defy splitting, and many more conundrums. My father's words when I asked for help with a problem were often, "Figure it out," meaning use your brain to think through the problem and solve it. His lessons I've never forgotten.

So thinking occurs on many levels, and there are many kinds of thinking. In this book, we discuss both critical and creative thinking and how to combine them in examining environmental challenges. Let's start with critical thinking. It's a topic many people have considered and have written about (see For Further Reading Section).

Critical Thinking

"Critical thinking is a process of identifying, synthesizing, analyzing, and/or evaluating information gathered from observation, direct experience, reflection, reading, communication with others or some combination of all of these."[3]

That definition is a mouthful, but critical thinking is a skill that needs revisiting and practicing a whole lot more these days. It is a process, along with creative thinking, that will help us make decisions about the challenges to our natural environment, and what actions we should take personally, and as part of a group, to meet these challenges. Critical thinking takes us beneath the surface, beyond what some radio or TV commentator is shouting about. Beyond internet articles that may or may not be accurate. Beyond information that is clearly *dis*information.

In our judgment, not enough people in today's society think critically. Sadly, many people have found it is far easier, and more convenient, to allow someone else to do their thinking for them. If that person comes up with what seems to make sense and appears to be in accord with what they believe and value, they sign on and go along with the ideas presented. In our judgment, that is a serious mistake. We need to stop and *think*. We each have the ability and even more importantly, in a democratic society, the responsibility for doing our own thinking—and doing it in depth.

Obstacles

- Too difficult to do overall.
- Lack of trust in science.
- Too difficult to sort out true and reliable information from disinformation or clearly false information.
- Too busy doing other things.

Creative Thinking

Creative thinking is defined as the creation of new ideas or approaches from something that already exists, or discovering something entirely new. It is sometimes defined as a nonlinear approach to solving a problem, examining a situation, or observing a societal challenge in a new and fresh way.

In 1983, when I was teaching at the University of Wisconsin–Madison, I came across a fascinating little book with a very nonacademic title: *A Whack on the Side of the Head.* In addition to teaching my regular courses, I also taught noncredit workshops on creative writing. At the time, when I began talking about creativity and creative thought, several students let me know that for them, you were either born with a creative bent or you were not. There's a sliver of truth to this argument, but only a sliver because each of us, in our estimation, has a creative self that is aching to be released.[4]

According to author Roger von Oech, creative thinking involves an imaginative phase and a practical phase. The imaginative phase results in new ideas and new approaches to doing things, while the practical phase involves examining, testing, and putting into practice the approaches that emerge from the imaginative phase. The first is the most difficult for many people; the second, although essential, sometimes prevents the first from happening.

To explain the situation in another way: the critical self, honed through years of education and experience, has helped us learn how to examine and make judgments about everything from books we read to a TV we might purchase—and much more of course. Most people are comfortable with being critical, but too often fail to allow their creative self to come out of the shadows and present new and innovative ideas—many perhaps never tried before. Unfortunately, the critical self often kicks in during the process of creative thinking—saying things in our minds such as "That's a dumb idea" or "Why'd you come up with that? It will never work." So, the creative self is often thwarted and prevented from providing us with new and innovative ideas and strategies for action.

Obstacles

- **Being Practical**. Ultimately, being practical is important. But screening every new idea, every new strategy will prevent creative thinking, and kill off ideas before they have a chance to be born. What on the surface may appear as a most impractical thought, may upon reflection and modification prove to be highly practicable. There is a time to be practical, but we must keep the idea of practicality in the background until several ideas are presented.

- **Fear of Being Wrong**. Most people do not want to be associated with "being wrong." We are a society that applauds and recognizes those who are "right" about things. Being wrong usually means minimally a reprimand or even worse—dismissal from a job, for example. But being wrong from time to time is often a necessary precursor to finding something that works, something that is right. Thomas Edison is claimed to have tried 1,800 light bulbs before he found one that would work. One thousand, eight hundred wrong answers before he discovered the right one.

- **One Correct Answer**. A common assumption held by many is that questions or problems have one correct answer. That may be the case in mathematics, but when thinking about combating threats to the environment, there may be several correct answers.

- **Discipline Blinders.** One of the negative aspects of our highly specialized society is specialists in a particular field often have difficulty seeing beyond their specialization. A plant pathologist, steeped in causes and cures for various plant maladies, may be able to easily see the potential of a certain chemical to cure a specific plant disease, but may not be able to see how it would cause havoc someplace else in nature.

Applying Critical and Creative Thinking

When dealing with most issues and problems, it is important to combine critical thinking with creative thinking. No matter if we are trying to figure out our understanding of climate change and its many ramifications to a more specific activity such as understanding the importance of cover crops to prevent soil erosion on farms.

Someone Who Combines Critical and Creative Thinking

- Knows how to separate facts from fiction and knows how to systematically assess the reliability of sources and the accuracy of their information.
- Avoids falling for the herd's perspective. Just because several people believe something is so, this person is forever skeptical and chooses instead to do their own independent analysis.
- Avoids following a charismatic charlatan who relies on emotion over facts.
- Is not afraid to question authorities.
- Avoids making hasty decisions because something "sounds right" or because it is the way something has always been done.

- Is open to new perspectives and ideas that may be different from the prevailing perspective.
- Is willing to question long-held personal beliefs when new evidence clearly contradicts a former position.
- Understands how science works, including its strengths and weaknesses.
- Is aware of the role that emotions can play in preventing, or seriously influencing, both critical thinking and creative thinking.
- Knows that storytelling, the power of the story, can get at what people are thinking about something and represents a form of communication that goes deeper than merely exchanging facts.

Outcomes of Combining Critical with Creative Thinking

- Solves a problem—from how to unclog a sink to what to do about increased pollution in waterways.
- Provides a foundation for an opinion—whom to vote for, developing a position on a national issue, and much more.
- Provides a foundation for an action plan.
- Results in learning something new—from discovering the meaning of quantum mechanics to identifying a new wildflower you have seen.
- Helps debunk false information—learning how to identify sources and the credibility of information. How to identify truth from lies.

Obstacle to Combining Critical and Creative Thinking

Combining critical and creative thinking is not always easy, nor can it be done quickly. In addition to the obstacles to critical and creative thinking mentioned earlier, a powerful block is our long-held, and often unknown, beliefs and biases. Our personal histories make a major contribution to our beliefs and values—they influence us far more than many of us realize. Where and when we were born, what our parents believed and valued, the nature of our early schooling, our relationship with our siblings and parents, and much more influence who we are today.

For example, I was born in the middle of the Great Depression on a small dairy farm in central Wisconsin. We had little in the way of the conveniences we take for granted today—no electricity, heating our home with woodstoves, no indoor plumbing, and elementary schooling in a one-room school with one teacher for all eight grades. In my early grades at the school, the schoolhouse had no electricity, never had indoor plumbing, and was heated with a woodstove. But I had family— mother, father, and twin brothers—and neighbors who stood ready to help each other at the drop of a hat. Very early in life, I learned what community meant—that we worked together, even though the farmers in our community were very different from one another with several ethnic backgrounds and religions represented, and we all got along. We had to because we depended on each other. No one had much money; several of our neighbors were barely surviving during those long and dreadful days of the Great Depression.

I also didn't realize how much the coming and going of the seasons affected how I saw the world. I wrote a book about winter titled *The Quiet Season*. In it, I wrote, "Winter has shaped me in ways that go deeper than I am even aware. . . . Living through a real winter—a northern winter— affects how we think, influences what we believe is important, and causes us to relate to other people in a particular way."[5] These experiences have had a great influence on me.

Until I was off to college, I didn't realize that most people's histories were not remotely similar to mine. By seeing who I was not, by getting to know my classmates, I became aware of who I was. This realization came into even sharper focus when I served in the army with young men from New York, Philadelphia, Chicago, and other big cities. From these young men, I learned something of their histories, what growing up in a large urban area was like, and how it was affecting what they believed and valued. Everyone's personal history has a substantial influence on who they are today, even though many years may have passed since they were kids.

Power of the Story

One of the first activities to do for those interested in becoming more adept at critical and creative thinking is to explore their personal stories. For many years I taught a writing workshop that I titled "Telling Your Story." I also wrote a book with that title. In these workshops, I helped people get in touch with their memories and their stories, and write them down. Here is what I wrote about the importance of personal stories:

> Stories help make us human; when we forget our stories, we forget who we are. Stories ground us, give us pleasure, and provide a sense of purpose in our lives. Stories help us recall the past while opening a window to the future. As new societal challenges emerge, revisiting the past stories of individuals and communities can offer insight to those making decisions about the future.
>
> Stories evoke feelings and lead the way to deeper thinking. By weaving facts into a story, we touch on feelings and move ourselves to think more deeply about what we are reading or hearing, especially as it relates to our own lives.[6]

One thing I tell all of my writing students is, "We don't know where we're going until we know where we've been."

It is important that we write our stories, for as we write them, we often uncover beliefs and values that may have been hidden from us. And these beliefs and values inevitably influence our thinking processes. I have written and published many of my personal stories. Reading them, it is clear that I have a bias toward country life, which I continue to hold. But in some of my stories, I uncovered some of my negative beliefs about city life and city living. I recognized that this was wrong, and I have tried to eliminate these beliefs from my thinking. I also discovered that leaving behind old beliefs is not as easy as one might think. I am convinced that we grieve the loss of long-held beliefs that we're ready to let go of. This grieving process may take minutes or months, as each of us grieves the loss of something in our own way. It is especially difficult if the belief we hold goes back to our childhood. My dad often told me when I was a kid that we should avoid swimming. He said, "Too much swimming saps the strength from your muscles." Later, I learned how wrong he was and that just the opposite is true. Swimming can strengthen one's muscles.

The Place for History

Too often these days, many people see an examination of the history of issues as a colossal waste of time. Writer Calvin Rutstrum said it well when he wrote, "Our eyes are focused forward. We seem to be inexorably bound up with the future, presuming that the new should hold the chief interest, the old holding just hollow criteria, although retrospection can hold future's guide."[7]

Almost every issue involving the natural environment has a history, from soil conservation (think the Dust Bowl of the 1930s) to land use issues (think zoning legislation that restricts certain activities on fragile lands), and of course many more. To do critical and creative thinking it is essential that we know the history of the issue, who was involved, when and where, and what strategies were used in solving the problem—if indeed the problem was solved.

But history can also be a deterrent to thinking, especially when the specter of "We've always done it this way," rears its ugly head. Critical and creative thinking requires that all strategies, those that have been followed in the past as well as those never before heard of or tried, are offered. Most of us know that just because a strategy worked in the past doesn't mean it will work in the present.

Combining Critical and Creative Thinking

Andrew Baker wrote this about the importance of combining critical with creative thinking:

> While critical thinking analyzes information and roots out the true nature and facets of problems, it is *creative* thinking that drives progress forward when it comes to solving these problems. Exceptional creative thinkers are people that invent new solutions to existing problems that do not rely on past or current solutions. They are the ones who invent solution C when everyone else is still arguing between A and B. Creative thinking skills involve using strategies to clear the mind so that our thoughts and ideas can transcend the current limitations of a problem and allow us to see beyond barriers that prevent new solutions from being found.[8]

CHAPTER 3

A Game Changer

*Combining Critical and
Creative Thinking*

Before looking at an approach for doing critical and creative thinking about environmental challenges, here is an example of what often occurs in our society today. Let's look in on Joe, who lives in a small Midwestern town and drives an old Ford pickup, which he has driven for fifteen years. The pickup, with 175,000 miles, is badly rusted and in need of about everything from new tires to a new engine. Joe really needs to buy a new vehicle. He discusses his situation with his brother-in law, Jim, who lives in Milwaukee, a brother-in-law Joe never liked, but puts up with because he's his wife's brother.

Jim says, "Why don't you consider buying an electric vehicle?"

"Aren't electric vehicles just another fad?" asks Joe, who has enjoyed his old pickup and looks forward to buying another similar one.

"Not a fad," says Jim. "It's what we need to do to protect the environment."

"So, what's wrong with the environment? If it needs protecting, let somebody else worry about that," Joe says. "What I need is a new truck."

"It won't be long before almost everyone will be driving electric vehicles," says Jim.

"What I'm looking for is a pickup that gets great gas mileage and doesn't need much maintenance," Joe says.

Jim laughs. "Then you should definitely buy an electric vehicle. Tell you what, why don't you do an electric vehicle search on the internet. You'll learn a lot about both their advantages and disadvantages."

"I'm not much for doing stuff on the internet—my wife looks at it a lot more than I do," offers Joe.

Nevertheless, once back home, Joe turns on the family computer and does a search for electric vehicles, as his brother-in-law suggested. First to come up is the website "Citizens for a Patriotic Society." He begins reading:

Electric vehicles are much in the news today. Don't believe what you read about them. They are a ploy to ruin the petroleum industry, which has so successfully provided the world with gasoline and much more. Electric vehicles sound great. They make little noise. You can't hear them coming, which makes them dangerous. Trying to cross the road when electric vehicles are present. Be careful or you'll be killed.

What if you're driving your electric vehicle and the battery goes dead? You can't walk to the nearest gas station and ask for a gallon of electricity. You'll need a tow truck.

The internet piece goes on for two more pages, disparaging electric vehicles at every turn, and applauding the petroleum industry. A final comment in the piece: *Concerned Americans will continue to drive gasoline-powered vehicles. It's the patriotic thing to do.*

Based on what he has just read, and the comments from his brother-in-law, Joe decides against purchasing an electric vehicle. He jumps in his worn-out pickup and drives to the nearest dealership in search of a new gas-burning vehicle.

Has our friend Joe practiced critical and creative thinking as he tries to decide what kind of vehicle he should purchase? Hardly. What he has acquired are opinions, not facts. And for these opinions, he only had two sources. His brother-in-law and an opinion piece on the internet. It would seem logical that Joe would want answers to questions such as: What is the price of electric vehicles compared to gasoline-powered vehicles? What is the expected life of an electric vehicle's battery and what does it cost to replace it? How long does it take to charge the battery on an electric vehicle? How safe are electric vehicles on the road? In what ways do electric vehicles contribute to solving environmental pollution problems? What negative effects do gasoline-powered vehicles have on the environment?

We'll come back to Joe and Jim in a bit. But let's shift to how we might use critical and creative thinking to address some of today's environmental challenges. I suggest four phases:

Phase One: **Examine Our Personal Beliefs and Values About the Environment**

Phase Two: **Identify and Research the Specific Environmental Problem You Wish to Pursue**

Phase Three: **Identify Possible Solutions to Solving the Problem(s) Identified**

Phase Four: **Select Action Steps and Take Action**

Here is an explanation of what is involved with each phase.

Phase One:
Examine Our Personal Beliefs and Values About the Environment

People hold a variety of beliefs and values related to the natural environment. Before moving forward with critical and creative thinking about specific environmental problems, examining those personal beliefs and values is a good place to begin. First some definitions. A belief is "confidence in the truth or existence of something not immediately susceptible to rigorous proof."[1] Values are the "importance, worth, or usefulness of something."[2]

For many people, once they bring their often-buried beliefs and values to the surface, they are open to change—open to accepting new ideas, new research, and new perspectives. And for those who continue to hold strong opinions, either for or against the natural environment, once they have examined their personal beliefs and values, often discarding those that are outdated, irrelevant, or just plain wrong-headed, they can move forward in making decisions with confidence in their position.

Examples of Beliefs and Values Related to the Natural Environment

With which of these beliefs do you agree/disagree?

- Facts about the natural environment are obtained by following the scientific method.
- Not everything there is to know about the natural environment can be obtained through science.
- We can come to know the natural environment through direct experience; for example, walking in the woods, watching a sunset, planting a garden, smelling a wildflower, feeling the rough bark of a bur oak tree.
- Knowledge about the natural environment can come from several sources: science (mathematics, chemistry, physics, etc.), social sciences (anthropology, archaeology, economics, sociology, etc.), art (writing, painting, sculpting, dance, etc.), and humanities (history, philosophy, religion, etc.).
- No one source of knowledge about the natural environment is more important than another.
- Knowing the natural environment goes beyond gathering facts about it.
- Some dimensions of the natural environment defy knowing.

- We can learn what needs to be known about the natural environment from writers, radio and TV commentators, and others who speak and write about it.

To which of these values do you subscribe?
- The natural environment is only of value in monetary terms.
- The natural environment has value beyond money.
- Elements of the natural environment—a wildflower, an insect, or a butterfly, for example—have value just because they exist.
- The natural environment both helps us see our place in the larger universe and helps us be humble.
- The natural environment can often have both economic and aesthetic value.
- The natural environment has negative value when it impedes economic development.

By examining and thinking about our responses to these statements, we can begin to uncover beliefs and values that we hold, and which, often without us even being aware of it, influence the position we take on many issues and influence us as we search for the facts.

By bringing beliefs and values to the surface, we can then decide if we wish to continue holding them or instead consider alternatives that would be more fitting to our present stage in life and the present state of the natural environment. Many unknown beliefs and values come from our childhood, the beliefs and values our parents held, and the beliefs and values that were prevalent in the community where we grew up. Beliefs about religion and politics generally have these roots. And so do our beliefs and values about the environment.

Here is a personal example of my examination of an environmental belief. On the home farm, my mother always cared for a flock of laying hens, usually around a hundred of them. Their eggs were one source of food for our family. My mother sold the extra eggs and used the money to buy groceries for the family as well as birthday and Christmas presents. This was her money. She was both proud of, and very protective of, her chicken flock. Any critter, such as a weasel, fox, or hawk—she called them "chicken hawks"—that would steal a chicken was the enemy. Thus, I grew up with the belief that any critter that potentially might raid the chicken house should be shot on sight, no matter where they might be seen. In fact, when a chicken hawk was observed—probably a red-tailed hawk—soaring over the farm, it was not uncommon to shoot at it with a rifle. And if one was killed, it was fastened to a fencepost near the chicken house—the belief was this evidence would dissuade other hawks from stealing chickens. I held these beliefs for many years, not thinking about them, and not bothering to examine their validity. Some years ago, when I learned more about hawks and other raptors, the story of my mother and the chicken hawk surfaced. I have since changed my mind about hawks and their importance to the natural environment.

Examining these basic belief positions can help us uncover from the deeper recesses of our minds the beliefs we hold about the natural environment. Reading widely will help provide information about what the natural environment really is; attending lectures, talking with other people, and experiencing nature directly are all approaches to help us first uncover our beliefs.

A technique I used with my graduate students at the University of Wisconsin–Madison was to ask them to develop a paper on what they definitively did not believe. I would say to them, "You can't know what you believe until you know what you don't believe." Some struggled with this assignment, for the tendency is to only listen to and read materials presented by those who agree with you. But examining positions opposed to our own not only helps us bring into focus beliefs we may not have known we have, but may offer us alternative beliefs—perhaps some that we may wish to adopt after careful examination.

Once we have examined our own beliefs and values about the environment, we are ready to move on to do critical and creative thinking about the world's current environmental problems.

Phase Two:
Identify and Research the Specific Environmental Problem You Wish to Pursue

Facts and Opinions

Once you have examined your personal beliefs about the environment and have defined the question or problem, it's time to look for further information. Research its history and present-day situation. For example, if your concern involves modern-day agriculture, what are the specific threats from various agricultural pursuits to the environment? Search multiple sources for information. Thoroughly check the accuracy of the information found, including validity and reliability. Check to make sure that the information is based on facts and not on unsupported opinions. Be on the alert for disinformation that is usually designed to mislead and misinform. Sometimes it is helpful to attempt to identify the beliefs that others hold about the situation.

As you search, and I strongly suggest researching several sources, examine each piece of information and determine if it is a fact or an opinion. Here are some tips on how to do that.

A fact is a truth about something that is supported with evidence. A fact can be verified. We can determine whether something is a fact by checking the evidence including verifying the source as credible (more about this below). Evidence may involve numbers, dates, stories, scientific publications, what you have witnessed, and more. Facts are necessary to bolster an argument, describe a situation, or support an opinion.

An opinion is a view or judgment formed about something that may or may not be based on facts. One's opinions can and should change based on a change in facts. An opinion stated by itself has little credence. An opinion must be expressed with the evidence used in supporting it.

Let's go back to the exchange between Joe and his brother-in-law. First, let's look at the beliefs that appear to undergird the exchange between Joe and Jim. First Joe. Joe believes that there is no credible alternative to gasoline- and diesel-powered vehicles, and that electric vehicles are a flash in the pan and will disappear shortly. He also believes that there is nothing wrong with the environment, and if there is, it is someone else's problem. He believes that vehicles like his old truck have served him well, and he sees no need to change to something different.

Jim has a different set of beliefs. He believes that the environment is in trouble and that human beings are responsible for a major part of the problem. He believes that electric vehicles are one way to help solve the environmental problem. He believes that human beings are a part of the environment rather than apart from it, and thus each person has a responsibility to help solve the problem.

So far, we see two quite different opinions emerging, neither one supported with facts. Jim, however, to his credit, suggests that Joe should gather additional information before he makes up his mind about what kind of vehicle he should buy to replace his decrepit pickup. He suggests Joe look on the internet for further information. Joe agrees to do this, but unfortunately he ends up with highly biased, unsupported information. Which leads us to the question, what are reliable sources of information?

Reliable Sources of Information

- **Personal experience.** Let's say that you have decided to grow a vegetable garden and live in an area where there are deer, wild turkeys, and other critters that enjoy feasting on vegetables. What can you do to keep these critters out of your garden? You ask around and get an assortment of suggestions ranging from scattering mothballs in your garden to putting up a twelve-foot-tall fence. You try the mothball idea. It doesn't work. You remember your days growing up on a farm and that when your father wanted to keep something in an enclosed area, he used an inexpensive two-wire electric fence. The top wire was about four feet from the ground, the bottom wire about twelve inches. You think that what keeps something in would also likely keep something out. You try it and it works. You've been using this approach to critical control for more than twenty years. What is more credible than twenty years of practical experience?

- **Scholarly, peer-reviewed articles or books**. Peer reviewed means that the article or book has been read and evaluated for its accuracy by persons in the writer's field of interest. To use the Joe and Jim story above, Jim might look for research that was done on the efficiency of electric vehicles by noted researchers who have had their work reviewed by their peers.
- **Trade or professional articles or books**. Authors of such materials are practitioners in the field or endeavor that is being discussed. Joe might look for factual information provided by the manufacturer of electric vehicles.
- **Magazine articles, books, and newspaper articles**. These are made available for a general audience, but have been written by professional journalists and reviewed by editors. A word of caution—these sources often contain both opinion and researched material.
- **Blogs and websites**. Some information is well researched with evidence presented. Some information is completely unreliable and should be avoided.[3]

Evaluating Accuracy of Information Sources

Here are some tips for determining the accuracy of information. Is the author identified? Is he or she an authority on the topic? What are the author's credentials? Is the author the original creator of the information? Has the author written other pieces on this topic? Does the author stand to gain financially from sharing the information or are they selling something?

Who is the publisher of the material? Is it a reputable organization or institution? Generally, universities and government agencies have more credibility than articles published by private industry (look for a .gov or .edu at the end of a URL), and they do not have a monetary interest in the research outcomes. Private industry generally does.

Misleading Information

Today, several information sources have one purpose: to mislead or misinform. These sources include TV, radio/podcasts, certain print publications, and internet/social media outlets. The purpose of these information sources is to try to convince you that what they have to say is right, and everything else written and spoken about the particular topic is wrong. These sources often have many followers. Just because many people believe something is accurate doesn't mean it is accurate. Critical and creative thinkers verify that every piece of information is accurate before accepting it.

The creative dimension of critical and creative thinking looks at problems and questions from a new perspective. The creative thinker is freed from the past and its various approaches to thinking of novel and often unorthodox solutions.

Creative Thinking Strategies

While most of the strategies discussed above are essential for both critical and creative thinking, the strategies that follow are specific to creative thinking. **Mind mapping** is one of the easiest of the creative thinking strategies. Start with a big sheet of paper. In the center of the sheet, make a circle, and in it write what you wish to focus on. Let's say you write,

"Things my family can do in nature." Then draw a line from the center circle, make another circle and write, for example, "Put up a bird feeder," and then draw another line and write another activity, and so on. Soon you will have a paper filled with activities, all relating to the center item.

Mind mapping can be used in many other ways as well. Let's say you and a group of friends are unhappy with sand mines coming to your area to provide special sand for the oil company's fracking activities. You have read about the potential air pollution problems and the related health risks, but you don't know how to proceed to either prevent a sand mine from coming, or make certain that the mining company follows all the rules and keeps environmental damage to a minimum or zero. In the center of the big sheet of paper draw a circle and in it write, "Combating sand mines." Then, with your friends, draw circles and write in activities that you can do.

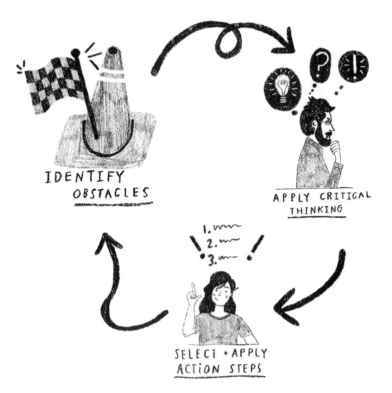

IDENTIFY OBSTACLES

APPLY CRITICAL THINKING

SELECT + APPLY ACTION STEPS

I have used this technique for many years with a variety of groups. Most people find it a useful way of discovering new ideas and new ways of thinking.

Brainstorming is another creative thinking activity that I have found useful. To do brainstorming with a group, write on a whiteboard the problem or question you want to pursue. Then, in rapid fashion, ask people to offer possible solutions to the problem. Using the example of a group opposed to the coming of a sand mine in their community, the group leader might write across the top of a whiteboard, "How to prevent a sand mine from coming to our community." Then ask for responses. Examples that might quickly be suggested: "Pass an ordinance preventing them from coming"; "Write letters to the editor"; "Organize a protest march"; "Get pollution evidence from other sand mines." Make sure that everyone in the group has an opportunity to speak. As group leader, do not comment on ideas suggested, even though some of them may sound impractical.

Another way to do brainstorming is to do it individually. Find a quiet place away from the blare of background noises. With a blank sheet of paper, write across the top the problem or question you want to explore, as was done with group brainstorming. Set a timer for ten minutes. Then write ideas, sometimes single words, that seem to have some relationship to the topic at hand. Write as quickly as you can. You will be surprised at what will appear on the paper in front of you.

Freewriting is a third activity to stretch your creative muscles. During freewriting, the writer doesn't stop writing, usually for a set period, and doesn't stop to edit or change what's on the page even if there are mistakes or if the ideas are unorganized. It's stream of consciousness writing.[4] Let your thoughts take you where they naturally gravitate.

Phase Three:
Identify Possible Solutions
to Solving the Problem(s) Identified

Be sure to include those solutions that might be creative and not considered previously. Examine the advantages and disadvantages of each potential solution.

Phase Four:
Select Action Steps and Take Action

Consider how much risk is associated with each action step. Are the action steps practical? After a specific action is taken, evaluate it and make corrections and adjustments where necessary.

Now we shift to examining specific environmental problems with suggestions as to how to apply critical and creative thinking to solve them.

CHAPTER 4

Climate Change

Finding Our House Is On Fire

Background

Here, we apply the lessons learned in the first part of the book to one of the most critical issues of the day. There is perhaps no hotter environmental topic than climate change, with signs of a planetary-scale crisis everywhere—intense rainfall and floods, soaring temperatures and drought, and wildfires.

"Climate change" and "global warming" are frequently used interchangeably, but they actually have different meanings. The National Aeronautics and Space Administration (NASA) provides clear definitions.

"Global warming" is the long-term warming of the planet. Global temperatures have been rising since the early twentieth century and more sharply since the late 1970s. "Worldwide, since 1880 the average surface temperature has risen about 1°C (about 2°F), relative to the mid-20th-century baseline (of 1951–1980). This is on top of about an additional 0.15°C of warming from between 1750 and 1880." While the term "climate change" includes global warming, it covers a broad range of planetary impacts including those affecting wildlife, agriculture and forests, other ecosystems, and human health.[1]

Satellite data and on-the-ground research, such as analysis of ice cores taken in Antarctica and Greenland, are used to track evidence of historical climate change. Other evidence of climate change can be found in tree rings, rock layers, ocean sediments, and the bleaching and dying off of coral reefs.[2]

How We Got Here

1600s—Coal begins replacing wood as a common fuel. In addition to being readily available, coal is energy dense, and it takes less coal than wood to produce the same amount of heat.[3]

1712—Thomas Newcomen, a British ironmonger, invents the first widely used steam engine. This invention helps launch the Industrial Revolution and contributes to larger scale use of coal.[4]

1800s—This period sees the invention of high-pressure boilers to power coal-fired engines on trains and steamboats.[5]

1824—The term "greenhouse effect" is first used by French physicist Joseph Fourier who writes, "The temperature [of the Earth] can be augmented by the interposition of the atmosphere, because heat in the state of light finds less resistance in penetrating the air, than in re-passing into the air when converted into non-luminous heat."[6]

1861—Physicist John Tyndall proves that water vapor, as well as other gases, are responsible for generating the greenhouse effect.[7]

1880s—NASA's Goddard Institute for Space Studies, the National Oceanic and Atmospheric Administration's (NOAA's) National Climatic Data Center, and the UK Meteorological Office's Hadley Centre—three records that track temperature—are born. At the time, people are not rattled by the suggestion that humans produce CO_2 and other gases that could collect in the atmosphere. In fact, in the 1890s, some welcomed the idea that the planet was warming.[8]

1896—Swedish chemist Svante Arrhenius determines that coal burning is contributing to the greenhouse effect.[9]

1900—Knut Angstrom, another Swedish scientist, discovers that "even at low concentrations in the atmosphere, CO_2 strongly absorbs parts of the infrared spectrum and can produce greenhouse warming."[10]

1908—The first mass-produced Model T Ford leaves the factory. By 1915, Ford is annually selling 500,000 Model Ts. Prior to 1900, there were fewer than 50,000 cars and trucks on the road. By 2000, the world saw more than 700 million vehicles traveling its streets and highways.[11]

1927—"Carbon emissions from fossil fuel burning and industry reach one billion tonnes per year." (A tonne is 1,000 kg or 2,204.6 lbs.)[12]

1930—The world population reaches 2 billion. This population growth, along with economic growth, leads to increases in human-produced greenhouse gases.

1958—Data supports that global warming is not only real but also that it is a serious concern. Scripps Institution geochemist Charles David Keeling begins measuring atmospheric CO_2 at Mauna Loa in Hawaii and in Antarctica. The project proves that CO_2 concentrations are rising. This rise is illustrated in what becomes known as the "Keeling Curve."[13]

1960—As computer climate modeling becomes more advanced, scientists predict possible results of rising CO_2 levels as evidenced by the Keeling Curve. The world population reaches 3 billion.

1970s—Although some believe global cooling and another ice age is potentially underway, most climate scientists predict global warming, not cooling.[14]

1980s—The early 1980s sees a big global temperature increase. Acid rain due to air pollution occurs in Europe and North America. The summer of 1988 is the hottest on record to date and witnesses pervasive drought and wildfires within the United States. The media and politicians start paying more attention to the serious environmental impacts of global warming. The public takes notice too.[15]

1989—The United Nations establishes the Intergovernmental Panel on Climate Change to provide "a scientific view of climate change and its political and economic impacts." Researchers warn of "severe heat waves, droughts and more powerful hurricanes" due to rising global temperatures.[16]

1997—The Kyoto Protocol, the first worldwide agreement to reduce greenhouse gases, calls for "reducing the emission of six greenhouse gases in forty-one countries plus the European Union" between 2008 and 2012 to "5.2 percent" below 1990 levels.[17]

2001—President George W. Bush announces that the United States will not implement the Kyoto Protocol, saying it is "fatally flawed in fundamental ways." He says the deal could hurt the US economy.[18]

2001—The UN's Intergovernmental Panel on Climate Change (IPCC) issues "its third report on climate change, saying that global warming, unprecedented since the end of the last ice age, is 'very likely,' with highly damaging future impacts."[19]

2006—Carbon emissions from fossil fuel burning and industry reach eight billion metric tons per year.[20] Al Gore debuts his film *An Inconvenient Truth*; Gore goes on to win the 2007 Nobel Peace Prize.[21]

2006—Climate change becomes highly politicized with skeptics arguing that predictions like those made by Gore and the IPCC are overblown.[22]

2012—Donald Trump, a climate change skeptic, on November 6, 2012, tweets: "The concept of global warming was created by and for the Chinese to make U.S. manufacturing non-competitive."[23]

2015—US President Barack Obama signs onto the Paris Climate Agreement. "In that agreement, 197 countries [pledge] to set targets for their own greenhouse gas cuts and to report their progress." The agreement strives to prevent a global temperature rise of 2°C (3.6°F), a warming increase that many scientists predict will "lead to increasing risk of more deadly heat waves, droughts, storms and rising global sea levels."[24]

2016—With Donald Trump's election as president, the United States declares it will withdraw from the Paris Treaty. President Trump states that he cannot "in good conscience support a deal that punishes the United States."[25]

2016—NASA and NOAA analysis finds that the Earth's 2016 surface temperatures are the warmest since modern record keeping began in 1880.[26]

2019—The UN Climate Action Summit suggests, "1.5°C is the socially, economically, politically and scientifically safe limit to global warming by the end of this century," and sets a 2050 deadline for achieving net zero emissions.[27]

2021—The world population reaches 7.9 billion.[28]

The Environmental Problem

The 2022 UN Intergovernmental Panel on Climate Change, a panel of more than 200 scientists, released a report suggesting "that if human-caused global warming isn't limited to just another couple tenths of a degree,"[29] some parts of the world will be uninhabitable.

"The cumulative scientific evidence is unequivocal: Climate change is a threat to human well-being and planetary health,"[30] says the report. By failing to act now to cut carbon emissions, the report warns, we "will miss a brief and rapidly closing window of opportunity to secure a livable and sustainable future for all."[31]

The report is clear—staving off the worst-case scenarios will require not only local but also global transformational change.

NOAA has historically tracked the planet's rising average surface temperature, with most of the warming occurring in the past forty years, and the warmest occurring during the seven most recent years (2016 to 2022). But rising temperature is just one indicator that climate change is occurring.[32]

During the last century, fossil fuel burning has led to an increased concentration of atmospheric CO_2, as has—although to a lesser extent—clearing land for industry and agriculture, as well as other human endeavors.[33]

According to NOAA

- Since the beginning of the Industrial Revolution, surface ocean waters have risen by approximately 30 percent.[34]
- "Sea levels have risen about eight inches worldwide during the last 100 years."[35]

In the Midwest, the Great Lakes—containing 84 percent of North America's surface fresh water, and a drinking water source for 40 million people—have been strongly affected by the negative effects of climate change. The EPA reports that warming waters enhance the growth of blue-green and toxic algae in the Great Lakes. Some coastal communities are forced to close beaches during algal blooms; they create health risks for humans along with their furry friends who like to take a dip on a hot day. Heavy rainfall flushes agricultural runoff from surrounding farmlands and feedlots into the Great Lakes. This runoff, which contains phosphorus, causes algae blooms and provides food for invasive species such as Asian carp, a species of growing concern in the Great Lakes that outcompetes native fish for food and habitat.

Homeowners along some shorelines are faced with a stench from rotting algae in hot summer months, impacting their aesthetics and property values.

Rising temperatures also curtail ice cover, which leaves shorelines pounded by wave action and more vulnerable to erosion. Extreme heat stresses farm animals and causes declines in meat, milk, and egg production.[36]

With global warming, forests face more frequent droughts, wildfires, and insect outbreaks. Many tree species, along with forest pests, are expected to shift their ranges northward.[37]

"Warming waters are expected to reduce the abundance of many cold-water fish species, including brook trout, lake trout, and whitefish."[38]

NASA Cites These Gases as Contributing to Global Warming

- **Water vapor.** "Water vapor is the most abundant greenhouse gas, but because the warming ocean increases the amount of it in our atmosphere, it is not a direct cause of climate change. Rather, as other forcings (like carbon dioxide) change global temperatures, water vapor in the atmosphere responds, amplifying climate change already in motion. Water vapor increases as Earth's climate warms. Clouds and precipitation (rain or snow) also respond to temperature changes and can be important feedback mechanisms as well."

- **Carbon dioxide (CO_2).** "A very important component of the atmosphere, carbon dioxide (CO_2) is released through natural processes (like volcanic eruptions) and human activities, like burning fossil fuels and deforestation. Human activities have increased the amount of CO_2 in the atmosphere by 50% since the Industrial Revolution began (1750). This sharp rise in CO_2 is the most important climate change driver over the last century."

- **Methane.** "Like many atmospheric gases, methane comes from both natural and human-caused sources. Methane comes from plant-matter breakdown in wetlands and is also released from landfills and rice farming. Livestock animals emit methane from their digestion and manure. Leaks from fossil fuel production and transportation are another major source of methane, and natural gas is 70% to 90% methane. As a single molecule, methane is a far more effective greenhouse gas than carbon dioxide but is much less common in the atmosphere. The amount of methane in our atmosphere has more than doubled since pre-industrial times."

- **Nitrous oxide.** "A potent greenhouse gas produced by farming practices, nitrous oxide is released during commercial and organic fertilizer production and use. Nitrous oxide also comes from burning fossil fuels and burning vegetation and has increased by 18% in the last 100 years."

- **Chlorofluorocarbons (CFCs).** "These chemical compounds do not exist in nature—they are entirely of industrial origin. They were used as refrigerants, solvents (a substance that dissolves others), and spray-can propellants. An international agreement, known as the Montreal Protocol, now regulates CFCs because they damage the ozone layer. Despite this, emissions of some types of CFCs spiked for about five years due to violations of the international agreement. Once members of the agreement called for immediate action and better enforcement, emissions dropped sharply starting in 2018."[39]

In 2021, just as some people were starting to climb their way out of COVID-19 pandemic confinement, other disasters were waiting for them outside.

- The temperature climbed to 130°F on July 9, 2021, in Death Valley, California, coming close to the hottest temperature ever recorded on the globe—134°F in 1913 at the same place.[40]
- More than 50 million people woke up to heat alerts stretching from northern Washington state to the Arizona–Mexico border.[41]
- Grand Junction, Colorado, set an all-time temperature record of 107°F on July 9, 2021, and Las Vegas tied its all-time high of 117°F degrees on July 10.[42]
- Eight states faced their hottest June on record: Arizona, California, Idaho, Massachusetts, Nevada, New Hampshire, Rhode Island, and Utah.[43]
- "Extreme Heat Cooked Mussels, Clams and Other Shellfish Alive on Beaches in Western Canada," shouted a headline on CNN.com.[44]

In fact, the nightly news and front pages of newspapers and news websites in 2021 were flush with stories of tragedy directly tied to climate change—flooding from hurricanes with names as friendly as Ida, intensifying drought conditions across portions of the Midwest and central Plains, and record flash floods across parts of New Jersey and New York, which resulted in fatalities when families became trapped in their basement apartments. Wildfires in Sequoia & Kings Canyon

National Park in California threatened some of the oldest and largest sequoia trees in the world.

According to the National Centers for Environmental Information (NCEI), which tracks and evaluates climate events in the United States and globally that have great economic and societal impacts, the United States has sustained "308 weather and climate disasters since 1980 where overall damages/costs reached or exceeded $1 billion (including CPI adjustment to 2021). The total cost of these 308 events exceeds $2.085 trillion. In 2021 there were at least 18 weather/climate disaster events with losses exceeding $1 billion each to affect the United States."[45]

Overall, these events led to "the deaths of 538 people and had significant economic effects on the areas impacted." Sources for the NCEI report included the US Army Corps of Engineers, National Weather Service, Federal Emergency Management Agency, US Department of Agriculture, National Interagency Fire Center, individual state emergency management agencies, state and regional climate centers, media reports, and the insurance industry.[46]

Climate change also impacts labor and sectors such as transportation, retail, tourism, and human health. A loss of snow negatively impacts ski resorts, for example. Road construction cannot be completed in floodwaters. Warm, moist environments breed mosquitos and make outdoor recreation miserable. Love bird-watching? An Audubon finding suggests that "two-thirds of American birds—389 of 604 bird species—are threatened with extinction from climate change."[47]

Crop losses, which threaten not only farmers' livelihoods but also food security across the globe, can be caused by climate extremes such as floods and droughts. In addition, both weeds and pests can flourish under increased temperatures and precipitation.

Obstacles

- Environmental science exists in a hyperpolitical context. Conflicts and concerns within society over natural resources and waste management can influence funding that is available for research. In applied environmental science, research is often funded by its relevance to special interest groups that have certain political backing.
- Pick your battle. Many voters have ranked global warming as a lower priority than issues such as health care costs or crime.[48]
- People bring their values, assumptions, and political biases to the table when discussing climate. Some distrust "evidence," and there is disagreement on what we come to accept as true.
- Because environmental problems, such as climate change, are interdisciplinary and include human health, the environment, economics, and more, a variety of critical and creative thinking approaches may be necessary.
- Climate change is complex. Sophisticated climate models and a data dump of decades of charts and graphs are not for everyone.

Applying Critical and Creative Thinking

Our concerns about the environmental impacts of climate change need to include concerns for humankind around the world.

"The course of climatization—the process by which climate change will transform society—will play out in the coming years in every corner of society. Whether it leads to a more resilient world or exacerbates the worst elements of society depends on whether we adjust or just stumble through."[49]

How do we heal? How do we move on from an obsession with wealth accumulation—one that pits multibillionaires in a contest to get Captain Kirk to space first (Jeff Bezos won that one)—to a society based on community care and connecting with nature? Unlike for COVID-19, there is no vaccine to stop the spread of global warming. But, we join many others in believing there is a cure to what is ailing the planet. We believe that we are the problem, but also the answer.

Key to critical thinking is to cite the facts and recognize that people are causing global warming.

Studies published in peer-reviewed scientific journals "show that 97 percent or more of actively publishing climate scientists agree that climate-warming trends over the past century are extremely likely due to human activities."[50]

This conclusion is echoed by the Statement on Climate Change from eighteen scientific associations: "Observations throughout the world make it clear that climate change is occurring, and rigorous scientific research demonstrates that the greenhouse gases emitted by human activities are the primary driver."[51]

The California Governor's Office for Planning and Research "lists the nearly 200 worldwide scientific organizations that hold the position that climate change has been caused by human action."[52]

In its *Fifth Assessment Report*, the Intergovernmental Panel on Climate Change, comprised "of 1,300 independent scientific experts, concluded there's a more than '95 percent probability that human activities over the past 50 years have warmed our planet.'" The panel's full Summary for Policymakers report is online.[53]

And the public is increasingly on board. According to the Pew Research Center, "Overall, about half of Americans (49 [percent]) say human activity contributes a great deal to climate change, and another 30 [percent] say human actions have some role in climate change. Two in ten (20 percent) believe human activity plays not too much or no role at all in climate change."[54]

If a wide range of research shows that a majority of the scientific community and the general public believe climate change is real and that action should be taken, why isn't more being done?

This is where creative thinking—and even moving past thinking to actually creating—can help us. If the facts aren't enough to convince the naysayers that climate change is happening and that humans are contributing to it, perhaps we move past presenting numbers in reports and turn to artists to show, instead of telling us, that there is a crisis. What stories do their paintings, poetry, and pottery tell us about place in nature? Instead of shocking with statistics, maybe when talking about climate change, we instead need to think about inspiring action.

The flexibility of art—that it appears in so many forms—means that there is likely a style for everyone. Art takes many shapes and can be shared.

Social media, such as Instagram, provides a worldwide audience for photographic evidence of climate change. Drones allow for time-stamped aerial photography of icebergs and lakes that are shrinking, along with islands being swallowed up by rising sea levels.

A chilling example of climate change art was the 2018 installation "Ice Watch," which featured massive ice blocks taken from icebergs in Greenland, shipped to Europe, and then placed in a ring of six blocks in London, where they stayed until they melted away.

In describing the exhibit, artist Olafur Eliasson (who created "Ice Watch" along with geologist Minik Rosing) wrote,

> The blocks of glacial ice await your arrival. Put your hand on the ice, listen to it, smell it, look at it—and witness the ecological changes our world is undergoing. Feelings of distance and disconnect hold us back, make us grow numb and passive. I hope that Ice Watch arouses feelings of proximity, presence, and relevance, of narratives that you can identify with and that make us all engage. We must recognise that together we have the power to take individual actions and to push for systemic change. Come touch the Greenland ice sheet and be touched by it. Let's transform climate knowledge into climate action.[55]

Those melting Artic ice blocks, displayed for all to see as they passed on the street, were examples of creativity born from creative thinking about a critical climate issue. If one listened carefully, one could hear the ice crack as it melted. Those blocks told a story of climate change in a way that a spreadsheet never could.

Art celebrates beauty but is also an instigator for change and raises awareness of social issues while documenting them over time. Art challenges us to see the world in new ways and to sometimes present what might otherwise be indescribable with words, charts, and spreadsheets. Art can make data accessible. It can change the way we think.

Selecting and Applying Action Steps

Now that we know more about the history and complex issues leading to and surrounding climate change and its impacts, what can we do to reduce our contributions to global warming and climate change? There are no easy answers, there's no magic bullet. Instead, we offer some suggestions on where to start taking action that is informed by thinking critically and creatively.

- Offer a viable alternative to fossil fuels, including cleaner energy sources. But there must be specifics attached to that plan—what will this cost? How long will it take? What technology is needed? Recognize that solar or wind might not be as effective in one area as another. Give examples where use of alternative energy sources has worked on a large scale—and on smaller ones.

- Emphasize "American made" and energy independence, with less reliance on foreign energy. Freedom from foreign influence or dependence is a concept most Americans will get behind. Coal miner jobs, for example, "have been declining for years, from a high of 92,000 workers in 2011 to just 52,804 in 2019," according to the Energy Information Administration.[56] Clean energy requires workers; for example, to design and install a wind turbine, you need engineers, electricians, truckers, and mechanics. Provide training to and incentivize the next generation of clean energy workers.

- Find balance. We believe that with creative thinking and action, there is a way to achieve both environmental health and economic growth.
- Appeal to personal self-interest. Inform the conversation with ways one can assert personal choice and save money or avoid waste.
- Make a local connection—create pride of place. According to the Pew Research Center, "most Americans today (62 percent) say that climate change is affecting their local community either a great deal or some." And

the vast majority of this group says long periods of unusually hot weather (79 percent of those asked or 49 percent of all U.S. adults) represent a major local impact of climate change. They also say major effects include severe weather such as floods and intense storms (70 percent), harm to animal wildlife and their habitats (69 percent), damage to forests and plant life (67 percent) or droughts and water shortages (64 percent). More frequent wildfires and rising sea levels that erode beaches and shorelines also are cited by equal percentages (56 percent of those asked) as major impacts to their local communities.[57]

- Put a face on it. Do you know someone who is fighting climate change–influenced disasters such as fires or flooding? Much of the burden falls on the back of frontline workers, such as firefighters and first responders, who are responding to disasters. Talk to these workers.

- Find trusted and independent sources. Highlight diverse coalitions and collaborations in support of conservation efforts. To overcome the partisan divide on the issue of climate change, seek out those with more credibility to various audiences. These include faith-based organizations, public health agencies, veterans, and business and social leaders.

- Speak to people's bank account balances. Greater storm risks lead to rising insurance rates. Someone has to pay for the devastation. "In the United States, insurance payout for natural catastrophes in 2017 and 2018 climbed to $291 billion total, the highest ever for any consecutive two-year period."[58] Grocery costs increase when severe weather destroys crops.

- Learn about Indigenous communities who are not contributing significant amounts of carbon emissions, but who are still affected.

- Make health a highlight of your conversation. Earlier blooms of grasses and other plants mean more and earlier pollen in the air, triggering asthma and allergy symptoms. An increase in some bug populations creates another health risk. According to the US Centers for Disease Control and Prevention (CDC), cases of diseases carried by ticks, mosquitos, and fleas tripled in the United States between 2004 and 2016.[59] Deer ticks, for example, are expanding their range as warmer temperatures make previously uninhabitable areas welcome respites for the Lyme disease carriers. They are living through the winters in places that previously had been too cold. Heat-related illnesses are on the rise. According to a recent CDC study, one in four Americans with chronic obstructive pulmonary disease are nonsmokers. One culprit for both this and the rise in other breathing ailments, scientists suggest, could be rising levels of ground ozone from automobile exhaust and factory emissions.[60]

- Think about climate change as a human rights and environmental justice issue. This change in thinking requires us to be curious and to better understand others who might live in very different social structures, cultures, and geographic regions.
- Buy and grow locally, which helps eliminate the need to transport food over long distances.
- Work from home when possible. COVID-19 shows that humans can react to a crisis and adapt. For many, the commute to work shifted from cars and buses to walking to the dining room in slippers instead. People proved they could work from home, and with that carbon emissions from cars decreased.
- Stress impacts of climate change on future generations and express the link to the long-term well-being of families and leaving a legacy of stewardship.
- Look to youth. Sometimes young people are the most creative thinkers, and they should be both seen and heard.

An example of youth impact is Greta Thunberg, a Swedish teen and internationally known climate activist who earned fame by challenging world leaders to act to mitigate climate change. In August 2018, Thunberg stood in front of the Swedish Parliament with a sign: "School Strike for Climate."

Her protest generated supporters from around the world, and by "November 2018, over 17,000 students in 24 countries were participating in climate strikes. Thunberg was nominated for the Nobel Peace Prize in 2019."[61] She participated in the United Nations Climate Summit in New York City in August 2019, taking a boat across the Atlantic instead of flying to reduce her carbon footprint.

Thunberg, who has more than 5 million followers on Twitter (@GretaThunberg), tweeted on July 3, 2021: "June 2021 was the hottest June ever recorded in my hometown Stockholm by a large margin. The second hottest June was in 2020. The third in 2019. Am I sensing a pattern here? Nah, probably just another coincidence."

Then, on July 7, 2021, Thunberg tweeted a headline from *The Guardian*, "The Climate Crisis Will Create Two Classes: Those Who Can Flee, and Those Who Cannot."[62]

One way to help climate naysayers accept the science is to share the benefits that go along with tackling climate change. These include economic development, such as job creation, with the rise of new and clean technologies.

Another starting point for having a conversation about climate change is to evaluate how you know what you know. Has your opinion been based on experience, research, or perhaps storytelling? Sometimes presenting an environmental problem in the form of a story or putting a face on the problem can be more impactful than regurgitating data and citing statistics, especially when tackling an issue as complex and polarizing as climate change.

Take the story of the tortoise and the hare. Or in this case, the stories of the snowshoe hare and the sea turtle.

In some areas of the United States, snowshoe hares are in danger. They are no longer invisible in the winter against the snow. These hares stand out to predators in areas that should be snow covered but are instead mud, making them easy prey.[63]

In other areas of the world, turtles are struggling to fight the rising tide of climate impacts. During a trip to Edisto Island in South Carolina, I visited a state park. The ranger there led a sea turtle nest survey, and we were informed that for every 1,000 sea turtle eggs laid, only about one turtle reaches adulthood. Raccoons that raid nests are part of the problem. But nesting habitat is also destroyed as water levels rise and stronger storms erode beaches. "Warming oceans will change ocean currents, potentially introducing sea turtles to new predators and destroying coral reef habitat."[64]

"If I'm trying to change somebody who disagrees—I choose not to be holier-than-thou," says Jane Goodall. "You've got to reach the heart. And I do that through storytelling."[65]

CHAPTER 5

Agriculture

Growing from a Complex History,
Facing an Uncertain Future

Background

It is impossible to discuss today's environmental problems in this country without considering agriculture. Few industries are as closely tied to the environment as agriculture. And few industries are more closely related to the future of this country than agriculture, for it is farms and farmers that produce the vast majority of food consumed in the United States.

As is true of most industries, agriculture has seen vast changes from the days of the subsistence farmer, who, with a team of oxen, plowed up a few acres of land to feed his animals and his family. From pioneer days to 1935, farm numbers in the United States continued to climb. In 1880, about 4 million farms could be found scattered across the country. In 1935, the number of farms peaked, with 6.8 million noted. Then a steady decline occurred, which continues to this day. In 2021, only a few more than 2 million farms still exist.[1]

In 1830, 91 percent of the US population of 12,866,020 lived on farms and in small villages. America was essentially a rural country.[2] It wasn't until 1920 that more than half the US population lived in urban areas. The country was built on its farms and villages. As recently as 1940, 43 percent of the country's population of 131,669,275 was considered rural.[3]

In the United States, farmers from the early settlement days to World War II essentially farmed as their fathers and grandfathers farmed. As a personal example, before the end of World War II on the farm where I was born and raised, which was 160 acres, we had no electricity, no indoor plumbing, and we farmed with horses. My elementary education was provided by a one-room country school, which was about a mile from home. My brothers and I walked as there were no school buses.

With a large garden and a small orchard, a small flock of laying hens, plus butchering a hog in the fall, we were nearly self-sufficient for our own food. My mother took care of both the garden and the chicken flock, selling excess eggs and using the money to buy our clothes and basic food supplies such as salt, sugar, coffee, and flour.

Most of the farms in our community, located west of Wild Rose in central Wisconsin, were similar to ours. They ranged from 80 acres to 160 acres. All were dairy farms, with farmers milking from a dozen to as many as fifteen or twenty cows by hand. Most farms also had a few hogs and a small chicken flock. Neighbors depended on each other at harvest time when they all gathered for threshing, silo filling, corn shredding, and wood sawing bees, gatherings where neighbors got together to help each other saw firewood. These events, in addition to the hard work involved, were also social events where there was considerable storytelling and practical jokes, as well as wonderful meals.

My father valued the land, second only to his family. He knew the value of replenishing the soil's fertility by spreading manure from our cattle and other livestock. He rotated his crops so the land had a chance to recover. He did not till steep hillsides but instead left them in grass. He pastured his cattle and hogs during the nonwinter months. He planted forage crops, such as alfalfa and clover, that helped restore the nitrogen content in the soil.

The Great Depression of the 1930s, with low prices and several years of drought, followed by four years of war with rationing and other restrictions, challenged the country's farmers. Most of them hung on, however, including my parents. Starting after World War II great changes began occurring. In 1936, almost 90 percent of US farmers did not have electricity. With the passage of the Rural Electrification Act of 1936, by 1950 nearly 80 percent of US farmers had electric power.[4] With electricity, electric motors replaced windmills and gasoline engines for pumping water. Electric motors powered milking machines and replaced milking cows by hand. Electric motors powered elevators that moved hay bale to the barn lofts and grain into granaries.

By 1950, almost every farmer had a tractor, although many kept their draft horses as well. With tractors came newly developed forage harvesters, eliminating traditional corn binders. In addition, there were tractor-powered grain combines, which eliminated the need for grain binders, shocking grain by hand, and for the threshing machine to move from farm to farm. Silo filling, threshing, and corn shredding bees disappeared. With tractors and electricity, plus new disease-resistant and higher-yielding crop varieties, farmers could grow more crops and raise more livestock with less human power than ever before.

Smaller family farms began disappearing. As noted earlier, the United States had 6.8 million farms in 1935. By 2020, only 2.02 million farms remained. New technology, including improved crop and animal genetics, chemicals, and equipment, induced farms to become larger and the number of farms to decrease dramatically.[5]

A 2005 US Department of Agriculture report summarized what happened in twentieth-century agriculture:

> American agriculture and rural life underwent a tremendous transformation in the 20th century. Early 20th century agriculture was labor intensive, and it took place on a large number of small,

diversified farms in rural areas where more than half of the U.S. population lived. These farms employed close to half of the U.S. workforce, along with 22 million work animals, and produced an average of five different commodities. The agricultural sector of the 21st century, on the other hand, is concentrated on a small number of large, specialized farms in rural areas where less than a fourth of the U.S. population lives. These highly productive and mechanized farms employ a tiny share of U.S. workers and use 5 million tractors in place of the horses and mules of earlier days. As a result of this transformation, U.S. agriculture has become increasingly efficient and has contributed to the overall growth of the U.S. economy.[6]

The Environmental Problem

As agriculture in the twenty-first century became increasingly enamored with the industrial model, and the belief that "bigger is better," environmental concerns were often pushed aside in favor of economic considerations. Here, we will look at two industrial-size agricultural activities, monoculture crop farming and industrial-size animal operations, and show how critical and creative thinking can be applied to addressing the environmental problems that have emerged as a result of their adoption. First, we will look at industrial-size monoculture crop farming.

Monoculture Crop Farming

Industrial-size monoculture crop farming means growing only one crop, usually as much or more than 1,000 acres, on a given area of farmland. This type of agriculture includes large acreages of single crops such as corn, wheat, and soybeans, as well as canning crops such as sweet corn, peas, green beans, and cucumbers.

During the years that I was growing up on a farm (late 1930s and 1940s), we practiced what was called "diversified agriculture"—the opposite of monoculture farming. We had about 20 milk cows, 25 hogs, and 100 chickens, plus 2 draft horses. Each year we harvested about twenty acres of hay for the cows, another twenty acres of oats, and twenty acres of corn for home animal feed. Additionally, especially during the 1930s, my dad grew twenty acres of potatoes, which was a cash crop, meaning most of the crop was sold off the farm. By the 1940s, we also grew up to a half acre of cucumbers, and by the late 1940s, a half acre or more of green beans, both sold as cash crops. A quarter acre or so of strawberries also provided my mother with a little additional money. A large vegetable garden provided almost all of the vegetables for the family. A small orchard provided us with apples. With the exception of the hay crops, such as alfalfa and the clovers, no crop was on the same land for more than one year.

Since World War II, however, and especially during the past several decades, industrial-size monoculture crop farming has become the norm in much of the United States.

Industrial-size monoculture crop farming has several potential negative environmental impacts:

- Can lead to increased pest problems. Because of the lack of biological and genetic diversity, weeds and insect pests can spread faster in monocultures. There are no natural defenses or barriers to naturally repel plants to stop pest infestations.[7]

- Decreases biodiversity. By definition, monoculture farming eliminates the diversity of plants on a piece of land. Because of the loss of diversity, and the natural resistance provided by plants and organisms, farmers end up adding increasing amounts of fertilizers, pesticides, and water to their monoculture fields.

- Degrades soil. Missing in monoculture fields are the soil organisms that are crucial for soil structure and composition. These naturally occurring organisms, such as bacteria, fungi, and earthworms, break down organic matter, distribute nutrients, and remove parasites and other harmful organisms.

- Compacts soil. Running heavy machinery over a monoculture field compacts the soil and alters its structure, making it difficult for rain and nutrients to move beneath the soil's surface.

- Potential runoff pollutes surrounding area. "No matter how much water and fertilizers are applied, these soils are not able to take them in and utilize them efficiently for crop growth. The result is increased runoff of concentrated nutrients that pollute surrounding areas, while cultivated lands yield less and less . . . excessive fertilizers and pesticides not only change the chemistry of surface water, they also leak into the groundwater and drinking wells."[8]

- Requires more water. Farmers often need to irrigate monoculture crops more than other crops; at the same time, the capacity of soils to absorb the water and retain moisture diminishes. Water is wasted, simply running off or quickly evaporating from the unshielded surface. Monoculture vegetable farming, such as on droughty soils found in the Central Sands area of central Wisconsin, requires massive amounts of irrigation water to succeed. After a multiyear study in the area, the Wisconsin Department of Natural Resources (DNR) found that high-capacity irrigation wells were responsible for lower water levels in lakes in the Central Sands region. As a result of the study, the Wisconsin DNR recommended the creation of a water use district to manage the resource in central Wisconsin. "According to the report, a water district could balance conservation and economic objectives through the use of practices such as metering, voluntary water-use reductions, incentives, withdrawal limits, land retirement, and more efficient irrigation practices."[9]

- Kills pollinators such as bees. Pesticides used to control pests on monoculture crops also kill bees and other pollinators. Researchers at the University of Wisconsin–Madison studied the decline of bumblebees (an important pollinator species). "As farmers cultivated more land and began to grow fewer crops over the past 150 years, most bumble bee species became rarer in Midwestern states."[10] Many crops, especially fruit crops, such as apples, cherries, and cranberries, need bees to set fruit.

- Heavy dependence on fossil fuels. Monoculture farms follow an industrial model. The crops are grown to be sold off the farm, requiring extensive fossil fuels to haul the crop from field to market, which may be thousands of miles away. And of course, extensive amounts of fossil fuels are required to plant and harvest the crop.

- GMOs (genetically modified organisms). The majority of the corn and soybeans planted in monoculture industrial-size farming in the United States are GMOs. This means that the genetics of the plant have been modified so that it is resistant to insects, insecticides, herbicides (such as Roundup), or often all. GMO crops were first introduced in the United States in the mid-1990s.[11] GMO crops have several advantages (weed and pest control requiring less effort is a major one), but they also have disadvantages. When weed and insect pests are chemically controlled, some always survive. These survivors pass on their resistance traits to the pests that follow them. With each crop planted, more survival pests, sometimes known as superweeds, appear. Heavier applications of herbicides are usually not the answer for controlling them. A debate continues: on the one hand, GMOs have dramatically increased the efficiency of food production. On the other, there are questions about the high use of chemicals in GMO crop production causing concern about the safety of genetically modified food.[12]

Obstacles

Obstacles to confronting the environmental impacts of monoculture crop farming include the following beliefs that many farmers hold.

- To survive, a crop farmer must have a large operation.
- Economics is more important than the environment.
- I own the land and can use it any way I want.

- Research and technology will eventually solve any problems that emerge (for example control of superweeds and new insect pests).
- Concern for the environment is somebody else's problem.

Applying Critical and Creative Thinking

Once it's decided that the environmental effects of industrial-size mono-cropping need remediation, a deep dive into critical and creative thinking about the issue should occur. One result of this effort includes a series of alternative cropping approaches.

Sustainable Agriculture

A growing number of farmers and agricultural scientists are moving toward a more sustainable form of farming, meaning sustainable environmentally, economically, and socially. This form of agriculture advocates building and maintaining healthy soil, managing water resources responsibly, promoting biodiversity, and minimizing air and water pollution.

Sustainable Agricultural Practices

- Integrating livestock and crops. In industrial agriculture, such as dairy CAFOs (concentrated animal feeding operations; see pp. 67–78), the cows often live a distance from where their feed is produced. Research shows that the integration of crops and animals on a farm can lead to more efficient and profitable farms.
- Rotating crops ensures healthier soil and improved pest control.
- Planting cover crops when the land would otherwise be bare helps build soil health, prevents soil erosion, keeps weeds in check, and replenishes soil nutrients.
- Reducing soil tillage (inserting seeds directly in undisturbed soil) helps to reduce erosion and improves soil health.
- Minimizing use of chemical pesticides by following a practice called "integrated pest management." This involves mechanical and biological controls for crop pests, and minimizes use of chemical pesticides.[13]

Organic Farming

Organic farming is a system of management and agricultural production that combines biodiversity with practices that preserve natural resources. Organic farming is defined as "an agricultural system that uses ecologically based pest controls and biological fertilizers, derived largely from animal and plant wastes and nitrogen-fixing cover-crops."[14]

Organic Farming Principles

- No use of synthetic fertilizers or pesticides.
- No GMOs.
- All organic waste is recycled.
- Crops are rotated to improve soil regeneration.
- Pests are controlled with biological agents.
- Animal welfare includes grazing outdoors.
- The environment is respected, and natural resources are preserved.
- Biodiversity is practiced by cultivating various crops and raising more than one animal species.

Farm-to-Table Movement

Following this approach, farmers—generally small-acreage or organic farmers—sell their produce directly to nearby restaurants, grocery stores, and sometimes directly to customers who enjoy the fresh food as well as appreciate the decreased environmental impact of food grown nearby. The benefits of farm-to-table farming are several:

- Peak freshness. Local produce ripens on the plant and is often harvested only hours before it appears on a restaurant plate. In this way, the customer enjoys the highest amount of nutrients and flavor. Food that has to travel a great distance, say from California to the Midwest, is often picked well before it is ripe, ripening on its way to stores and food distributors.

- Environmentally friendly. Hundreds of gallons of fuel are saved if the food is available nearby rather than traveling great distances. Thus air pollution is avoided from burning fossil fuels, and contribution to climate change is reduced.
- Local jobs. Farm-to-table supports local farms as well as local restaurants, coffee shops, and other eateries. There is reduced reliance on large, profit-driven corporations that may focus on maximum production over animal health and welfare.
- Variety. Local seasonal foods provide a wide variety of choices that are grown and harvested during their optimal growing seasons.
- Learning opportunity. Farm-to-table can help urban people learn about the food resources in their community such as local vineyards, vegetable gardeners, and organic farmers.[15]

Small Space Farming

Community gardens, which are usually found near urban centers and offer space for consumers to grow their own food, are one alternative. A backyard garden, similar to the idea of Victory Gardens that were popular during World War II is still another. On a personal note, both Natasha and I have backyard gardens. Natasha and her husband, Steve, grow strawberries, rhubarb, garlic, chives, basil, tomatoes, hot peppers, broccoli, kohlrabi, cabbage, green beans, lettuce, zucchini, yellow summer squash, cucumbers, kale, and sunflowers. During the summer of 2021, my backyard garden (only four-by-eight feet and surrounded by a fence to keep out rabbits) grew tomatoes, cucumbers, pole beans, lettuce, and a few annual flowers.

My family also has a large garden at the family farm in central Wisconsin, which Natasha and Steve manage. In 2021, the garden included five rows of sweet corn; a row of zucchini and winter squash; a row of broccoli, green and red cabbage, and kohlrabi; two rows of snap beans; a row of carrots and beets; a row of lettuce, kale, spinach, and radishes; six long rows of potatoes; six rows of tomatoes; two rows of cucumbers;

a row of gourds; and a row of sunflowers. Three Apps families enjoy the vegetables from this garden. Some of the produce, such as potatoes and squash, last well into the winter, while tomatoes and sweet corn are preserved for use throughout the winter months.

Small-acreage farmers grow vegetables and meat for a local market. They usually sell their products at local farmers' markets, to local groceries, or directly to customers. Some small-acreage farmers are part of the community supported agriculture (CSA) movement whereby farmers and consumers are connected by the consumer subscribing to the farmer's harvest. Others are part of the farm-to-table movement.

Aquaculture

Aquaculture includes two basic types: marine and freshwater. Marine seafood, which is generally grown in confined areas in an ocean, includes shrimp, clams, oysters, and mussels. Freshwater food, grown in ponds, includes bass, catfish, tilapia, trout, and salmon.

Commonly referred to as fish farming, aquaculture has a long history. It is believed that the Chinese were practicing aquaculture as early as 2000 BC. The practice soon spread to other countries in Asia and Europe, and later to the United States. Popular fish farms in the United States produce salmon and trout. Some advantages and disadvantages of aquaculture follow.

Advantages

- Reliable source. Provides a good, dependable source of seafood. With aquaculture, there is decreased demand on freshwater and ocean fisheries, which are being rapidly depleted.
- Easily established. Artificial ponds and tanks can be established almost anywhere. They need not be close to natural bodies of water.
- Jobs. Aquaculture requires considerable labor for its operations, thus providing employment opportunities in regions where it is established.

- Wild species protection. By taking stress off ocean and freshwater fishing, endangered species can be protected and allowed to replenish themselves naturally.
- Nutrition. Provides a substantial amount of nutritious food, often in areas where other types of agriculture may be challenged.

Disadvantages

- Invasive species may be introduced into the region where aquaculture is practiced.
- Water pollution may result if fish are allowed to stay in the same body of water for an extended period, and that water finds its way to natural lakes, rivers, streams, and drinking water sources.[16]

Vertical Hydroponic Farming

Different from traditional farming where crops are grown in soil and are heavily dependent on climate and weather, vertical hydroponic farming crops are grown indoors. No soil is needed. Crops are planted in trays or towers and fed with water that includes the specific nutrients required by each plant type. Within the growing house, with its vertical walls, conditions of light, heat, and humidity are carefully controlled for each crop. Hydroponic gardening uses a sterile growing medium; thus weed, disease, and pest problems can be minimized. Crops, such as corn, squash, pumpkins, and soybeans, and fruit crops such as oranges, apples, and cherries, do not lend themselves to this type of agriculture. Common crops for vertical hydroponic farming include lettuce, spinach, strawberries, bell peppers, and herbs.

With this system, these crops can be grown year-round, and use far less space than needed for traditional farming. "The complete lack of soil and closely controlled microclimates for each plant answer concerns relating to climate change and soil degradation. In fact, an acre and a half of vertical hydroponic farming can produce 500 tons of healthy, leafy greens spread over 11–13 harvests each year."[17]

Aeroponic Farming

Aeroponic farming, similar to hydroponic farming, allows farmers to avoid risks related to weather and pests. Because crops are grown stacked inside a building, only minimal acreage is necessary. Instead of soil, aeroponic farming crops are suspended in a fine mist of water and nutrients. A pump, connected to a timer, sprays aerated nutrient solution on the roots periodically. This process saves 95 percent of the water compared to farming with soil and saves 40 percent of the water compared to hydroponic farming, where the crop roots are submerged in water. Aeroponic farming also controls lighting, making sure that plants get the light they need, and only the light they need, which improves efficiency and reduces energy costs. Common crops grown include lettuce, strawberries, cherry tomatoes, mint, and basil.

An example of aeroponic farming is an indoor farm called Nebullam located in the Iowa State University Research Park. The indoor farm features walls of lettuce, microgreens, and tomato plants. The roots of the vegetables are sprayed with a nutrient-rich mist, which, in this case, allows for 98 percent less water needed compared to growing these vegetables in fields. The growing environment is entirely controlled, thus there is no need for pesticides. Originally, Nebullam distributed produce to local grocers and restaurants. With the COVID-19 pandemic closing down most restaurants, the company began delivering their produce directly to consumers at their homes.[18]

Vertical farming practices could lead to extremely efficient forms of food production and can be set up in the middle of urban areas, thus reducing the need for transportation and storage.[19]

Selecting and Applying Action Steps

Thinking both critically and creatively, the next phase is to consider which alternatives to industrial-size monoculture crop production could work for a given community or region, including deciding how to apply them. Each of the alternatives is examined in terms of its advantages and disadvantages, keeping in mind that environmental impact is a major concern. One way to begin this examination is to reflect on the beliefs and values that appear to undergird the alternatives identified.

Possible beliefs and values that emerge may include:

- Food can be produced without using soil (aquaculture, hydro-ponic, and aeroponic farming).
- Food can be produced on the land without harming it (organic farming, farm-to-table, small-acreage gardening, sustainable agriculture, and home gardening).
- Bigger is not better for many reasons, but especially because it is harmful to the environment.

Action steps may include writing letters to the editor of local newspapers, participating in radio call-in shows, sponsoring community meetings to discuss the problem, having booths at community events such as at farmers' markets to discuss the situation with attendees, support-ing small-acreage farmers, buying food that is locally grown, supporting restaurants that participate in farm-to-table programs, spreading the word on social media, creating a podcast, sharing on an online neighborhood listserv such as NextDoor, and writing articles for the local newspaper. Local lawmakers can also be contacted, especially concerning issues such as overuse of irrigation water.

Proactively, concerned people can participate in community gar-dens, grow their own gardens in their backyards, or sign up with a farmer who is a part of the CSA movement.

Industrial-Size Animal Operations

Let's now turn to an examination of the second agricultural activity: industrial-size animal operations and their environmental impacts. Spe-cifically, I have chosen to examine industrial-size dairy operations. The same approach can be applied to examining environmental impacts of poultry, beef, and hog industrial-size operations. I chose my home state, Wisconsin, as an example. Other major dairy-producing states would likely have similar experiences, including California, Idaho, New York, and Texas. Along with Wisconsin, these states collectively produce more than 50 percent of the milk supply in the United States.

Background

I was a farm boy who grew up milking cows by hand, and our dairy herd numbered about fifteen to twenty. I have trouble wrapping my mind around the idea that nowadays, a dairy herd might number several thousand milk cows. Over the past few years, several articles and stories in the media have examined the environmental impact of these large dairy farms. But first, what is an industrial-size dairy operation? The EPA refers to large-scale animal operations as concentrated animal feeding operations (CAFOs). (When I first saw those letters it reminded me of an urban gang, such as the Mafia. I wondered what the EPA was doing messing with the criminal underworld.) The letters stand for concentrated animal feeding operations, an agricultural enterprise where animals are kept and raised in a confined situation. A CAFO is further defined as having more than 1,000 animal units, with each unit equivalent to 1,000 pounds of live weight. This translates to 1,000 head of beef cattle, 700 dairy cows, 2,500 swine weighing more than 55 pounds each, 125,000 broiler chickens, or 82,000 laying hens or pullets confined on site for more than forty-five days per year. CAFOs are regulated by the EPA under the Clean Water Act of 1972.[20]

Traveling around southern and especially northeastern Wisconsin, it is easy to spot long steel buildings with canvas curtains on the sides that can be raised and lowered. Inside, dairy cattle are organized in two long rows eating in mangers with a driveway between them. Outside are long rows of what look like white dog houses; inside each one is a little calf. Usually there is also a tanker truck or two near the buildings used to transport the milk from what may be five thousand or more dairy cows at the farm. These are dairy CAFOs.

In Wisconsin, CAFOs are also regulated by Wisconsin's DNR. Each CAFO is required to have a manure storage unit—an open-air lagoon with enough capacity to hold six months' worth of manure. Once the manure spends several months in a lagoon, its smell seems to intensify (for more on the environmental impact of manure, see p. 70). A couple of years ago, I happened to drive on a road just after one of the enormous CAFO manure spreaders accidentally dumped a goodly amount of its load on the road. Unknowingly, I drove over it, and the smell that engulfed my car told me what I had done wrong. I immediately drove to a car wash, much to the unhappiness of the carwash employees who said they had never seen such a smelly car. I had to drive through the car wash twice and still some manure clung to the car's underpinning.

Wisconsin's dairy CAFOs are relatively new to the state, having their beginnings in California. By 2021, Wisconsin had 289 dairy CAFOs with a large concentration of them in northeastern Wisconsin. In that year, the state had 6,804 dairy farms, down from 14,158 as recently as 2007. But the number of CAFOs in Wisconsin—90 percent are dairy—has risen each year since 2005.[21]

Why the increase in these large dairy operations? What immediately became evident to me was that dairy farming was responding to the "get bigger or get out" mentality that has swept the business and industrial world for the past several decades. It could be called the "industrialization of farming." Small- and medium-size farms went out of business by the hundreds, while the larger ones grew ever larger.

Contributing to the "get bigger" philosophy is the fact that dairy CAFOs are more efficient than smaller dairy operations—more milk can be produced per animal, at less cost. Ultimately, this efficiency means that grocery prices for dairy products, such as milk, cheese, yogurt, and ice cream, will be reasonably priced. Another plus, the farm owner doesn't have to milk cows twice or three times a day, 365 days a year, but can hire workers to do it. Thus, a CAFO can be an important employer in the community. Still another advantage is that the CAFO operator can concentrate on one agricultural endeavor, in this case dairy, without having to become proficient in several, which was the case for the diversified farm owner. On diversified farms, not only did farmers have to be proficient in working with several different animal species, such as dairy cows, hogs, and chickens, but they also needed to be proficient in growing several different kinds of agronomic crops such as alfalfa, clovers, and pasture grasses along with corn, oats, and perhaps soybeans. Some CAFOs do grow much of their feed on land that they rent or own.

The Environmental Problem

The environmental impacts of dairy CAFOs cannot be underestimated. The major impact revolves around manure. Because the cows are confined in buildings year-round, all of the manure they produce must be dealt with in one way or another. In smaller-scale dairy operations, cattle are out on pasture for a major part of the year, and thus there is no accumulation of manure.

A mature dairy cow weighing 1,400 pounds can generate about 14 gallons (about 120 pounds) of feces and urine per day. Cities use sewage treatment systems to process human sewage. As mentioned, for CAFOs, most farmers store a mix of cow feces and urine in a lagoon. Usually twice a year they spread this material across crop fields.[22]

In Wisconsin's Kewaunee County, Mark Borchardt, a USDA microbiologist, researched private wells and discovered dangerous materials from fecal matter entering the wells, most of it coming from industrial-size dairy farms. During research he conducted in 2017, Borchardt discovered that more than 60 percent of the wells sampled in Kewaunee County were contaminated with fecal material. The risk factor for a well was determined by how close the well sampled was to a manure storage pit; wells as far away as three miles were at risk of being contaminated. In 2017, there were about 270 manure pits in the county. Kewaunee County has fractured bedrock, common in northeastern Wisconsin, which allows for relatively easy contamination of subsurface water. Similar research in southwestern Wisconsin showed that 42 percent of the wells in Iowa, Grant, and Lafayette Counties were contaminated.[23]

Will Cushman, writing for the *Wisconsin State Farmer,* stated,

> With so many animals housed in one place, CAFOs produce volumes of manure every day that must be dealt with somehow. The most common practice is to spread this manure on farm fields near a CAFO in the spring and fall. The manure helps return nutrients to the soil for crop production, but it also has the potential to seep into groundwater and wash into streams and lakes.[24]

According to the EPA, industrial-size agriculture is the leading contributor of pollutants to lakes, rivers, and reservoirs in the United States. Large animal waste lagoons, some as large as a football field, can also result in significant greenhouse gas emissions—a contributor to climate change.[25]

A second study conducted by Tucker Burch, a USDA agricultural engineer, reported that the number-one factor for gastrointestinal illness in Kewaunee County was caused by drinking water from wells contaminated with cow manure. The study found that the main source of illness from private wells was caused by *Cryptosporidium*, which was estimated to cause 250 illnesses per year.[26]

Obstacles

The following are some of the obstacles we need to overcome if the environmental impacts of dairy CAFOs are to be confronted and corrected. They are some of the beliefs and values held by those advocating industrial-size dairy operations.

- Bigger is better.
- A dairy operation is similar to a factory where inputs and outputs are carefully considered.
- For the dairy industry to survive, the CAFO approach is necessary.
- A milk cow can be viewed as a machine.
- Economic considerations are more important than environmental ones.
- A cow is of value only in terms of its ability to produce quantities of milk.
- Land is of value only for its ability to support the CAFO structures and provide a place to spread the manure.

Applying Critical and Creative Thinking

The questions to be answered: What are the specific environmental impacts of dairy CAFOs? What are alternative ways of lessening or eliminating these environmental impacts? What action steps can be taken?

After doing the research on dairy CAFOs, and examining this research for accuracy and reliability, the critical and creative thinker steps back and reflects on the information, including how to confront beliefs and values that are foundational to CAFO dairy farming.

A Different Approach

As I continued doing critical and creative thinking about dairy CAFOs, I reread some of Aldo Leopold's work. Aldo Leopold, writing in the mid-1900s, was an early voice in calling for a different approach to agriculture than the one that was beginning to emerge, especially after World War II. In an essay titled, "The Land Ethic," Leopold wrote, "Quit thinking about decent land use as solely an economic problem. Examine each question in terms of what is ethically and esthetically right, as well as what is economically expedient. A thing is right when it tends to preserve the integrity, stability and beauty of the biotic community. It is wrong when it tends otherwise."[27]

Those subscribing to agrarianism also see a fundamental flaw with CAFOs. Wendell Berry, a Kentucky farmer and author, is a leading supporter of agrarianism. He writes about a fundamental difference between industrialism and agrarianism. "I believe that this contest between industrialism and agrarianism defines the most fundamental human difference, for it divides not just two nearly opposite concepts of agriculture and land use, but also two nearly opposite ways of understanding ourselves, our fellow creatures and our world." Berry goes on to say, "Because industrialism cannot understand living things except as machines, and can give them no value that is not utilitarian, it conceives of farming as forms of mining; it cannot use the land without abusing it. . . . Industrialism begins with technological invention. But agrarianism begins with givens: land, plants, animals, weather."[28]

So what are the ways we can lessen or eliminate the negative environmental impacts of dairy CAFOs? Solutions vary from shutting CAFOs down, to modifying them to lessen their negative environmental effects, to encouraging alternate approaching to dairy farming. Here is a brief description of the alternatives.

Manure Digesters

A partial remedy to the environmental impact of dairy CAFOs is to require them to process their manure with a manure digester. A manure digester uses an anaerobic (low or no oxygen) process to break down the components in manure. One of the by-products of the process is biogas, a mixture of methane and carbon dioxide and other gases, which can be used to produce heat, electricity, and fuel. Other byproducts of manure digestion can be used as fertilizer and animal bedding.[29]

> Digesters can substantially reduce odor and releases of methane, a powerful global-warming gas; can convert nitrogen into ammonium, a form more available to plants and less likely to be

carried away with runoff when the remaining waste solids are land-applied; can reduce fly infestation; and can reduce the oxygen-depletion capacity of the remaining waste although the liquid waste does still require additional treatment prior to release.[30]

Manure digesters are expensive and not fool-proof technologies. One way for farmers operating CAFOs to make digesters more affordable is to create a cooperative neighborhood digester.

Sustainable Agriculture

For those who believe that dairy CAFOs are not sustainable as a long-term agricultural activity, many are advocating an approach to agriculture that is sustainable environmentally, economically, and socially. This approach to farming advocates maintaining healthy soil, concern for biodiversity, and eliminating air and water pollution as much as possible (for more information on sustainable agriculture, see p. 61).

Grazing

The grazier group's approach to dairy farming challenges the idea that bigger is better and that technology will solve all problems associated with CAFOs. It is not a new idea; indeed, it is a very old idea. My father was a grazier, but he never called himself that, because that was the way every dairy farm operated in those days. Being a grazier means allowing dairy cattle to be out on pasture during warmer months. During the colder months, the cows are indoors and are fed stored hay, silage, and grain. Farmers who follow this approach disapprove of the CAFO approach with large numbers of cattle remaining confined inside all of the time in most instances. Graziers believe it is beneficial for cows to have regular access to the outdoors with fresh air and sunshine and for them to eat natural growing, unharvested crops for several months of the year. Another practical benefit is when these animals are not confined, they are naturally spreading their waste on the fields where they graze. There are obvious cost savings when cows find their own feed and naturally dispose of their waste.

Organic Farming

Organic farmers consider animal welfare as an important tenet of their activities, as they consider caring for the land having high priority. They believe that all animals should have the opportunity to spend time outside and should not be confined for long periods (for principles of organic farming, see p. 62).

Dairy Goats

Goat milk has become a popular alternative to cow's milk. It has been a popular drink internationally for years. Goat's milk is especially good for people who are lactose intolerant and must avoid cow's milk. Compared to the various nut-based "milks," goat's milk has more protein and is loaded with calcium and other minerals such as magnesium and potassium. Nut milks often have a watery consistency, while goat's milk is creamy with consistency compared to cow's milk. It has small, well-emulsified fat globules. The cream will stay in suspension for a longer period than cow's milk and thus does not need to be homogenized.[31] In addition, goat farmers generally operate smaller farms, which lessens their environmental impact and also keeps the animals' welfare in mind. And the fact that goats are significantly smaller than dairy cows means they produce less waste.

Farmer-Led Solutions

The Dairy Strong Sustainability Alliance, formed in 2016, changed its name and broadened its focus in 2021. "The Dairy Business Association and the Nature Conservancy originally organized the alliance in Wisconsin around the goal of helping dairy farmers make tangible improvements to the environment and other aspects of their farms." Today, they are known as Farmers for Sustainable Food. Their vision is "a sustainable food system in which farmers, their communities and the environment thrive." The alliance includes members of The Nature Conservancy, farmer food processor representatives, and food companies. Steve Richter, with The Nature Conservancy in Wisconsin, said, "Farmers for Sustainable Food's connections with stakeholders throughout the agriculture supply chain, their strong relationships with farmers, and their ability to create well-structured and well-run projects have complemented our efforts to provide science, technical support and funding to help farmers be successful."[32]

Plant-Based Milk

A more extreme alternative to CAFOs is for the milk-consuming public to move from dairy milk to plant-based milk. Interest in plant-based milk has been on the increase for several years, as dairy milk is not the popular drink it once was in the United States. In 1975, the average American drank 115 quarts of milk a year. By 1990, milk consumption dipped to 102 quarts, 92 quarts in 2000, and 66 quarts in 2020.[33]

Veganism is growing in popularity around the world. Veganism advocates eliminating all animal products from one's diet as a way to improve one's health, protect the environment, and protect animals. A vegan diet consists entirely of plant-based foods—no eggs, milk, or meat. The veganism movement is increasing in popularity as measured by the retail sales of vegan products such as nut-based milk, plant-based meats, and other such products.

At the dairy section of the grocery store recently, I saw nearly as many containers of plant-based milk as I did of dairy milk. I saw oat milk, soy milk, rice milk, and almond milk. Several reasons account for the shift from dairy milk to these various nondairy milk products, one being the sizable number of people who are lactose intolerant or experience discomfort from milk protein.[34]

Laboratory-Produced Dairy Products

A company called Perfect Day said this in their promotional materials, "Finally, an option for dairy lovers and plant-based fans alike. We've invented the world's first milk proteins made without animals, so you can enjoy the taste, texture, and nutrition of traditional dairy, produced sustainably and without the downsides of factory farming, lactose, hormones, or antibiotics." Perfect Day's products are different from plant-based dairy products, such as almond, coconut, and soy milk. Cow-free dairy products result from creating proteins that are genetically identical to animal proteins.[35] The company's protein product is made by altering sections of the DNA sequence of food-grade yeast, such that the microorganisms, once fed with certain nutrients, produce several key proteins found in milk.

Ryan Pandya, one of the founders of Perfect Day, says, "Developing this science is about more than creating a tasty vegan-friendly [milk product]. Cutting down on the number of cows is a way to make the food industry more sustainable. These animal-free 'milk' products use 98 percent less water and 65 percent less energy to create [milk products] than ones that use dairy cows."[36]

Selecting and Applying Action Steps

The next question in the critical and creative thinking process is to ask which of the various alternatives to dairy CAFOs merit action. The solutions to the dairy CAFO environmental problem range all the way from how to prevent one from coming to your neighborhood, and how to live with one that is already there, to alternative approaches to dairy CAFO farming. Questions to answer include: What are the advantages and disadvantages of each of the alternative solutions to the problem? What are the obstacles to choosing one of the alternatives?

Consider that, as a result of critical and creative thinking about the research that you and a group have done, you all agree that dairy

CAFOs are not sustainable as most of them operate today. If you agree that the negative environmental impacts outweigh their positive contributions to society, what can you do? Based on the evidence uncovered, which of the above solutions to the problem seems most doable, or is it a combination of several that would make the most sense? Then action steps must be considered. Organize community meetings, write opinion pieces for the local newspaper. Contact your local and state politicians to develop stricter rules for manure handling. In addition, boycott CAFO operations, letting the operators know you don't agree with what they're doing; support organic farmers and other small-scale operators, such as goat and dairy sheep farmers; support sustainable farming practitioners; and applaud graziers for what they are doing. When considering action steps, additional questions must be asked: What are the risks associated with a specific action step? Is the action step practical?

Agriculture was and is a critical component of the US economy. More than almost any other economic endeavor in the country, agriculture's tie with the land is extremely close. Because agriculture means food for almost all of us, it must be sustainable. And it must, we would argue, take into account its environmental relationships as well as its economic dimensions. Getting to sustainability requires some careful critical and creative thinking on the part of all citizens, not only those who have a direct connection to agriculture.

CHAPTER 6

Forests

*Discovering Deep Roots
and Branching Out*

Background

My first introduction to forestry came when I was maybe three or four years old, and my mother asked me to fetch some wood from the wood-shed for the wood-burning cookstove that sat on the west side of our kitchen. At the time we heated our farmhouse with woodstoves, kept our pump from freezing with a woodstove in the pumphouse, and kept our cash crop of potatoes from freezing in the potato cellar that was built on a hillside just west of the farmstead. I knew that every fall, my dad spent several days "making wood," as he called the process of going out into the twenty-acre woodlot just west of the farmstead. With our team of horses and a bobsled, Pa hauled the oak logs and branches to the farm-stead. A few days later the neighbors arrived; one of them owned a huge, screaming circle saw, which cut the logs into proper length blocks for woodstoves. Next, the blocks were split with a splitting maul to a size where they would fit into the various stoves. By the time I was twelve years old, Pa was teaching me how to split wood, as the task was called.

At age eleven, I joined a 4-H club, and I signed up for the forestry project. By virtue of being a forestry member, I received twenty-five pine seedlings that I was to grow out in a little nursery bed. I received instructions on how to build one, with Pa's help. And soon I had the little trees, only a few inches tall, growing in my little tree-nursery bed in back of the chicken house. After two years in the nursery, I transplanted the little trees in a row, just north of our woodlot. Today, so many years later, I drive by these pines that are now seventy-five to eighty feet tall, and I feel good about what I did when I was eleven years old. I also realize that my love for trees has continued ever since. In the fifty-five years that I've owned the farm I have now, we have planted trees every year. Sometimes only a few—one year 7,500. I run the farm as a tree farm, working closely with a professional consulting forester to make sure I follow sustainable forest practices.

The forests of the United States have a wide variety of uses, going back to the days when Indigenous Americans used the forests for everything from collecting maple syrup to using forest products for shelter. When the first English settlers arrived in the 1600s, forests were cleared for farming, logs were used for buildings, and forest products were exported to England. By the late 1700s, New England was exporting millions of board feet of pine boards as well as logs for ship masts. By the 1830s, Bangor, Maine, had become the world's largest shipping port for lumber. At the time, it was believed by many that the forests of the United States were never ending, that they would always be there. By the beginning of the Civil War in 1861, many areas of the Northeast were completely logged out. Loggers moved west, to the Great Lakes states of Michigan, Wisconsin, and Minnesota, where the logging industry boomed as cities, such as Chicago and St. Louis, grew rapidly and settlers moved west, creating a huge demand for wood products.[1]

By the late 1800s, it became evident that the country's forests were not a never-ending resource, as some people had claimed. In 1889, George Vanderbilt hired Gifford Pinchot to manage the forest at the Biltmore estate near Asheville, North Carolina. Pinchot was educated in Europe and was the first professional forester employed in this country. Congress passed the Forest Reserve Act in 1891, creating a reserve of 40 million acres of forestland in the United States. In 1897, Congress passed the Organic Act, which provided funds for the management of the Forest Reserve.[2]

The US Forest Service, a part of the US Department of Agriculture, was created in 1905, with Pinchot as its first head. Until World War II, the Forest Service focused on watershed protection, reforestation, and wildfire prevention and suppression. With abundant supplies of private timber, little harvesting of national forest timber took place. But following World War II, with an enormous building boom, the national forests were seen as a ready supply of building material. Oftentimes, there was clear-cutting of national forestland, which led to increased environmental concerns. By the 1960s and 1970s, several laws were passed to protect forests.[3]

A Wisconsin Case Study

The first lumber speculators from the East arriving in Wisconsin saw millions of acres of virgin white pine, rivers to transport the logs, and waterpower to drive sawmills. These speculators began arriving by the late 1830s, but the industry progressed slowly, primarily due to lack of skilled workers. New England and eastern Canada lumbermen had more years of experience with lumber production. Knowing this, Wisconsin logging companies began recruiting men from New Brunswick, Maine, New Hampshire, New York, and Pennsylvania to come to the Wisconsin pinery to provide the expertise necessary for success.

Wisconsin's logging industry experienced its first boom years from 1850 to 1856. The industry slowed after the economic panic of 1857, but by 1860 it was once more booming. Wisconsin's wheat production was at its peak in the 1860s, and prosperous farmers bought lumber to add buildings and make other improvements on their farms. The Homestead Act of 1862 made 160 acres of federally owned land available to settlers without charge other than a $10 fee, the only caveat being that the farmer had to live on the property for five years and show improvements—which included buildings—within five years.

The market for lumber outside Wisconsin was also increasing. The rapidly growing big cities of the Midwest were a major market. And with the expansion of railroads and water shipping, a market for Wisconsin lumber grew even more. As the Civil War raged, the demand for lumber from Wisconsin's pineries had never been greater.

Wisconsin's abundant pine forests served as an engine for the state's economic growth in several ways throughout the 1800s. The mighty white pine tree was a catalyst for other iconic industries that are woven indelibly throughout Wisconsin's culture and economy, such as wood products manufacturing and papermaking.

Wisconsin's lumber industry continued to prosper into the 1870s. For a time, it seemed there was no end in sight to the logging boom. But as the end of the nineteenth century approached, even the loggers, who once believed the supply of pine logs was limitless, began to see an end. Some of the logging firms moved on to establish operations in the northwestern United States or in the South. They left behind what came to be known as the "cutover," a vast region of stumps and slashings.[4]

The Environmental Problem

Using Wisconsin as a case study, as early as 1854, Increase Lapham, a leading Wisconsin scientist of his day, wrote about the importance of forests to the environment. He described how forests influenced the state's climate, rainfall, and soil and warned of the dire consequences of allowing the clear-cutting of the north to continue unabated. In Wisconsin, the state legislature began to consider that something had to be done. In 1867, the legislature passed a law creating a forestry commission.

In the commission's report, Lapham minced no words about the likely outcomes if clear-cutting Wisconsin's vast pineland continued. He wrote, "A country destitute of forests . . . is only suited to the conditions of a barbarous or semi-barbarous people. Deprive a people of the comforts and conveniences derived directly or indirectly from forest products, and they soon revert to barbarism. It is only where a due proportion between the cultivated land and the forests is maintained that man can attain his highest civilization."[5]

The forestry commission's recommendations were prescient, but few people paid attention to its findings. The peak years for logging in the pineland of Wisconsin, for example, were in the 1870s and 1890s. The logging industry had considerable political clout, and made sure that the boom continued. Logging increased in the state every year until about 1900, except for the years of economic recession, 1873 to 1878 and 1893 to 1897. During these boom years, few people seemed to care about logging's long-range environmental impacts.

Economist Frederick Merk wrote in 1916,

> In every new country the natural resources closest at hand are the first to be exploited. . . . The gifts of nature are transformed into capital as rapidly as possible, and the means thus accumulated form the basis for the development of other fields of industry. In Wisconsin lumber was one of the easiest to transmute the great northern forests into gold. . . . With reckless disregard for the future they wasted the gift of the ages.[6]

Much of what happened in Wisconsin occurred in other states where logging was a major historical activity. After 1890, the state's lumber industry began a slow decline. Even so, Wisconsin continued to rank among the top five states producing forest products for several more years. Farmers were encouraged to farm the cutover land. But as early as the 1920s, it was clear to Wisconsin lawmakers, lumber interests, and many of the farmers themselves that much of the former timberland in northern Wisconsin was best suited for forestry, not farming. Much of the cutover soil was too sandy, and the short growing season prevented certain crops, such as corn, from maturing before frost killed the plant. Some farmers stayed on their poor farms, but many did not. Tax delinquency spread across the northern cutover as farmers packed up and left, abandoning their land for others to worry about. The Great Depression of the 1930s further devastated the cutover farms, reducing their average income by 40 percent between the 1920s and 1939.

On some cutover lands, huge gullies formed because there was nothing to slow down a heavy rain. With slashings still abundant, fire continued to be a threat. In 1903 and 1905, the Wisconsin State Legislature passed comprehensive forestry laws, which provided for the creation of a Department of State Forestry to be controlled by a Board of State Forest Commissioners, creating the position of state superintendent of forests (later called state forester). This created forest experiment stations, allowing the state to accept grants of land for forestry purposes, which established a system of state forests, providing for the disposition of public lands, and setting aside 62,000 acres in Forest, Oneida, Vilas, and Iron Counties as a forest reserve.

The Wisconsin State Legislature passed the Forest Crop Law in 1927, which included the provision that a property owner with 160 or more acres (later reduced to 40 acres) could declare his or her land best suited for forestry. The owner would sign a fifty-year renewable contract with the state. Other than a fee of 10 cents per acre, the owner paid no property taxes. To aid the municipalities, the state paid them 10 cents per acre for all lands in the forest crop program. By the end of 1986, the

Forest Crop Law program had enrolled about 1.5 million acres of private forestland. The Managed Forest Law (MFL) succeeded the Forest Crop Law in 1986, and in 2017, more than 3 million acres of private forestland were enrolled under the MFL.

The area of forestland in Wisconsin in 2020 is approximately 17 million acres. More than 46 percent of the state's land area is forested, primarily in the north and west. In 2020, 57 percent of the Wisconsin forests were owned by individual landowners such as farmers, homeowners, hunting partners, and others. Thirty-two percent of Wisconsin's forested land was owned by federal, state, county, or tribal governments, and 11 percent was owned by private corporations.[7]

Climate change is also a major challenge to forests. Within the United States, it is expected that the temperature, on average, will increase by 3 to 10°F in the coming years. This will result in warmer winters and hotter summers. Some parts of the country will see increased rainfall, while other parts will see severe drought, and one effect of drought is the associated increase in wildfires. In 2020, as of late November, the United States had recorded 52,113 wildfires that burned 8,889,297 acres, with most of these fires occurring in the western and northwestern United States. This was more than twice the acreage of forestland burned in 2019.[8]

In addition to drought and the resulting wildfires, climate change will have other effects on forests. For example, with warmer temperatures and a longer growing season, some tree species are likely to move north, and some tree species that have done well with shorter growing seasons and colder winters may suffer with warmer temperatures.

Climate change can result in more insect outbreaks in forests and increased invasive species. This could further result in some tree species either dying out or shifting their range. For instance, in Colorado, the pine beetle has damaged thousands of acres of forestland. Spruce beetles have damaged several million acres in southern Alaska and western Colorado. Climate change will likely increase the number of insects invading forestland. With increases in temperatures, some insect species will develop faster and continue to move north.

Climate change will also result in extreme weather events, such as windstorms, which can raise havoc with a forest. An increase in ice storms and hurricanes will also negatively affect forests in the United States.[9]

Deforestation resulting from urban growth and development is a continuing problem in the United States. Clear-cutting of timber in some parts of the world, in order to grow more cultivated crops, is another problem. Deforestation adds more carbon dioxide to the atmosphere as felled trees release stored carbon when they are burned to make way for industrial-size agriculture. Deforestation results in loss of habitat for animals, birds, and insects.

Trees transpire water into the atmosphere, and without trees, the surrounding air is drier and hotter, making it more difficult for foliage and plants that depend on water and the shade from trees to survive. Tree roots also bind the soil and help prevent soil erosion, especially during heavy rainstorms.[10] Large forest areas, in addition to providing wood products, also provide recreation, health, and well-being opportunities, as well as biodiversity, which includes mammals, birds, fish, reptiles, plants, fungi, and microscopic organisms. Well-managed forests also protect freshwater resources.[11]

Healthy forests require management, and resources for doing this work are limited, especially for national and state forestlands. More research is needed, especially concerning the effects of climate change on forestlands and the control of invasive species.

Fragmentation is another challenge facing the nation's forests. Forest fragmentation means dividing large parcels of forestland into smaller parcels, some as small as an acre. Along with the smaller parcels come roads, clearing for homes, and corridors for utility lines. In Wisconsin, for years large timber and paper companies owned massive acreages of forestland. Toward the end of the 1900s, these large firms began selling their forestlands, which often were purchased by developers and subdivided into smaller parcels.[12]

Obstacles

Unfortunately, trees and forests are taken for granted by many people. "Why should we worry, trees can take care of themselves?" is often heard. Others see trees as only an economic resource—lumber and paper from them are necessary for a society to prosper. These are people who often fail to see that trees are necessary for a healthy environment, beyond their economic value. Still others see trees as a nuisance, especially some urban dwellers who hate having to rake leaves in the fall, and shudder to think that a tree might fall on their home.

Applying Critical and Creative Thinking

What threats do forests face and what can be done about these threats? After reflecting on the history and the environmental challenges facing forests, and checking on the accuracy of the information we have gathered, we next reflect on the beliefs and values that undergird the information we have.

Beliefs and Values About Trees and Forests

- Healthy forests are essential for a healthy environment.
- The value of a forest exceeds its monetary worth.
- Forests not only provide wood products but also offer homes to plants and wildlife.
- Historically, the value of forests ranged from being a nuisance because they had to be cleared for agriculture to being a high-priced resource for their building materials.
- Forests provide a "carbon sink" to help curb climate change.

After studying the information we have collected, and after reflecting on the associated beliefs and values, we move to considering possible solutions to the problem.

- **Sustainable Forest Management**. Practicing sustainable forest management emerges as one important solution. Sustainable management practices focus on caring for forests based on the best forestry science, which results in healthy forests now and into the future. Sustainable forestry is concerned with all segments of the forest—trees, smaller plants, soils, wildlife, and water. It also includes protecting forests from wildfire, pests, and diseases, and preserving forests that are unique or special. An important practice of sustainable forestry is to determine whether the forest has enough natural seeds and seedlings to make a future forest. Other practices include putting up a fence to exclude deer, controlling invasive plants, and removing some trees to allow more sunlight to reach the forest floor. Sustainable forestry practices also include protecting forest streams and wet areas as well as controlling pests, diseases, and wildfires.[13]

- **Forest Fragmentation**. Preventing forest fragmentation is another way to help ensure healthy forests. As forestland is divided into smaller and smaller pieces, what is left is clearly not a forest. One way to prevent fragmentation is to support the Forest Legacy Program (FLP, which was included in Congress's 1990 Farm Bill). The FLP conservation program is administered by the US Forest Service working in partnership with state agencies. Its purpose is to encourage the protection of privately owned forestland through conservation easement or land purchase. By providing economic incentives to landowners to keep their forests as forests, the FLP program encourages sustainable forest management as well as supports strong markets for forest products.[14]

Selecting and Applying Action Steps

No matter if we live in a rural or urban area, we can help support sustainable forestry practices by first recognizing and appreciating the value of forestland to everyone. The value of forests ranges from providing wood products to offering a place for recreational activities. We can also help support legislation, especially concerning national and state forests, to fund sustainable forestry practices.

Private forest owners should consider signing up for a sustainable forestry program, if offered in their state. In Wisconsin, the Managed Forest Law Program (FLP) is administered by the Wisconsin DNR. It is an incentive program that encourages sustainable forestry on private woodland. In exchange for following approved sustainable forestry practices, the landowner pays reduced property taxes.[15]

Some private forest owners might consider the FLP as a way to protect forestland for future generations. Everyone, whether living in an urban area or in the country, can take a stand against forest fragmentation, through the ballot box, letters to the editor, and community meetings, where the issue of dividing a nearby large forest into smaller pieces comes up. One simple thing to do is to regularly take a walk in the woods, leaving behind your cell phone. Stop and listen, smell, and feel, as well as see firsthand what goes on in a forest. Here is a living, vibrant place; so different from a nonwooded area. Such an experience can provide a foundation for the action steps outlined above, as well as steps you might think of that aren't mentioned here.

CHAPTER 7

Water

Riding a Wave of Complex Issues

Background

I first realized the critical importance of water when I was a little shaver growing up on a central Wisconsin farm in the 1930s and 1940s. During the mid- to late 1930s, in central Wisconsin, and in many other western and southwestern states, a multiyear drought resulted in sandstorms that filled the skies with dust, destroyed crops, obscured the sun, and made life miserable for urban and farm people alike. At the time, we had no electricity at our farm, and a windmill pumped our water from a deep well. When the wind blew, and the windmill turned, we had sufficient water for our small herd of milk cows, a few hogs, a pair of draft horses, and a flock of laying hens. And of course, for the family's personal use.

Then on a hot day in August, the wind quit blowing, and the windmill quit turning. Within a day, the stock tank that provided the water for the animals was empty, and the cattle and horses had nothing to drink.

There was no water for the hogs and the chickens. And we had no water for washing, drinking, or cooking.

I went to bed with the sound of cattle bellowing for water, a sound that I never forgot. The following morning, I heard my dad on the party line telephone calling Allen Davis, who lived a half mile north of us. Allen had a gasoline-operated pumpjack to pump his water. I heard my dad asking, "Can we come down to your place and get some water for our livestock?"

After breakfast, we loaded the steel-wheeled hay wagon with empty ten-gallon milk cans. There must have been at least a half dozen of them. We hitched the team to the wagon, and we were off to the Davis farm. Soon we were filling the milk cans with precious water. When we got home we dumped the water into the stock tank where the cattle crowded to get a drink. Later that day, I saw Pa reading the ads in a farm paper for a used gasoline engine pumpjack to replace the windmill that had served us so well over the years—when the wind was blowing.

Human beings, along with other living creatures, need water to survive. Up to 60 percent of the human body consists of water. Water regulates our body temperature; moistens tissues in our eyes, nose, and mouth; flushes waste from our body; acts as a shock absorber for the brain and spinal cord; lubricates joints; and helps dissolve minerals and nutrients to make them accessible to the body. Every day we lose water through our breath, perspiration, urine, and bowel movements. A person can live for more than a week without eating, but only for about three days without water.

Seventy-one percent of the Earth's surface is covered with water. Water is also available beneath the Earth's surface and in the air in the form of water vapor. Water on the Earth is a closed system; the water that was here many billions of years ago is the same water that is here today. Even with an abundance of water, only a fraction of it (about 0.3 percent) is fit for human consumption. The remaining 99.7 percent of the Earth's water is found in the oceans, icecaps, and water vapor in the air. Of the 0.3 percent, some of it is surface water, such as freshwater lakes, rivers, and streams, and more of it is underground in aquifers and soil moisture (groundwater). Surface and groundwater are available for human use.

Surface water provides about 80 percent of the water used by humans for drinking, bathing, and so forth as well as for irrigation. Groundwater makes up 98 percent of the Earth's fresh water, which is primarily found in aquifers. Both ground- and surface water are subject to pollution, making the water unfit for human consumption.[1]

Fresh water use can be classed as "direct" and "indirect." Direct use refers to drinking, cooking, washing, and perhaps gardening. Indirect use refers to the water that is used to manufacture goods that we consume or produce. This includes water used to grow crops; care for animals such as poultry, beef, and dairy cows; food we purchase at the grocery store; as well as almost every other item we use in our daily lives. Indirect use of water also includes water used to generate electricity.[2]

The natural environment is heavily dependent on water. Not only do humans need water for their survival, every other living plant, bird, and animal needs water for its survival. Wetlands including ponds, lakes, rivers, and streams are essential for waterbirds, from providing a nesting place to offering a place for the young to grow and thrive. Wetlands also draw migratory waterfowl, especially ducks and geese, providing a safe place for them to rest before they continue their long, sometimes more than thousand-mile migratory journey. Wetlands also help to filter water as it moves through the system. Aquatic plants and flowers provide food and shelter to a wide range of insects, frogs, reptiles, and mammals.[3]

The Environmental Problem

Water pollution has been and continues to be a major problem in this country and around the world. A simple definition of water pollution: when harmful substances such as chemicals, microorganisms, runoff from farmers' fields, and a host of other pollutants contaminate a stream, river, lake, ocean, or aquifer and make the water toxic to the environment and to humans.

For many years, people thought little about water pollution, often contributing to it directly. For instance, I remember as a kid stopping by the Wild Rose Cheese Factory and seeing a pipe filled with wastewater dumped directly into the Pine River that ran nearby. Indeed, many industries, such as cheese factories, purposely located on streams and rivers to provide a ready means of getting rid of wastewater. Another classic case of wastewater pollution is the paper mills located on the Fox River in Wisconsin. For years they dumped their wastewater that contained highly toxic PCBs (polychlorinated biphenyls) directly into the river, which drained into Green Bay and then into Lake Michigan. A cleanup project eventually removed PCBs from the river.[4]

Examples of Water Pollution

Groundwater Pollution—groundwater is found in aquifers under the ground. When rain falls and seeps into the ground, it eventually adds to the aquifer. Some 40 percent of Americans depend on groundwater for their water supply, which is pumped to the surface from the aquifer. Groundwater is easily polluted from pesticides, fertilizers, waste leached from septic systems, waste leached from industrial-size dairy operations' holding lagoons, and similar sources, making the water unsafe for human consumption.

Surface Water Pollution—surface water is found in our oceans, ponds, lakes, and rivers. Unfortunately, a large percentage of our rivers, streams, and lakes are polluted and unfit for drinking, swimming, and fishing. The leading cause of pollution comes from nutrient pollution such as nitrates and phosphorus, much of it from farm manure and fertilizer runoff. Other pollutants are municipal and industrial waste discharges. Some 80 percent of ocean pollution originates on land, including contaminants such as runoff of chemicals, heavy metals, and nutrients coming from farms, factories, and cities via streams and rivers that empty into bays and estuaries and then into the ocean. Plastic is the most prolific pollutant in the ocean, blown by the wind or washed in via storm drains and sewers, for example—or even dumped illegally.[5] Unfortunately, too often an ocean is polluted by oil spills, some small and many large.

Point Source Pollution—this refers to contamination that originates from a single source. Examples include wastewater discharged legally or illegally by a manufacturer or oil refinery, leaking septic systems, as well as chemical and oil spills.

Nonpoint Source Pollution—this is pollution occurring from several sources, including agricultural and stormwater runoff. Nonpoint source pollution is the leading cause of water pollution in US waters. It is often difficult to regulate because there is no one identifiable source.[6]

Clean Water Act

In 1948, the US Congress passed the first Federal Water Pollution Act. With the environmental movement in full swing in the 1960s and 1970s, Congress passed a sweeping amendment to the act in 1972. The amended act became commonly known as the Clean Water Act (CWA). The 1972 amendments to the act included the following:

- "Established the basic structure for regulating pollutant discharges into the waters of the United States.
- Gave the US Environmental Protection Agency (EPA) the authority to implement pollution control programs such as setting wastewater standards for industry.

- Maintained existing requirements to set water quality standards for all contaminants in surface waters.
- Made it unlawful for any person to discharge any pollutant from a point source into navigable waters, unless a permit was obtained under its provisions.
- Funded the construction of sewage treatment plants under the construction grants program.
- Recognized the need for planning to address the critical problems posed by nonpoint source pollution."[7]

The Clean Water Act has become an important tool in caring for the nation's water, but there remain concerns to be addressed.

Remaining Concerns

As discussed, agriculture continues to be a major polluter of rivers, streams, lakes, and groundwater. Agricultural pollution consists of fertilizer and pesticide runoff, as well as runoff from animal waste. Excess nutrients in water, especially nitrogen and phosphorus, can cause blue-green algae blooms that are harmful to humans and wildlife. Untreated wastewater is another contributor to water pollution, as well as continued illegal dumping of wastewater by a few industries.[8]

Not only has water pollution been a problem for the nation's water supply, overuse of water has also become an issue. In central Wisconsin, irrigation of droughty soils to support monoculture industrial-size crop farming has drawn down the aquifer to the point of drying up nearby lakes.

More than a billion people worldwide lack ready access to water, and a total of 2.7 billion find water scarce for at least one month of the year. Many worldwide water systems that provide water for a growing human population and keep ecosystems thriving, such as rivers, lakes, and aquifers, are becoming too polluted to use or are drying up. Wetlands throughout the world are disappearing. Agriculture continues to consume more water than any other source. Climate change around the world is causing shortages and droughts in some areas and floods in others. "At the current consumption rate, this situation will only get worse. By 2025, two-thirds of the world's population may face water shortages. And ecosystems around the world will suffer even more."[9]

Looking ahead, the world is destined to have insufficient quantities of fresh water. Climate change will have a major influence on the distribution of the water, which will result in people either having not enough water or too much. At this writing, most of the western United States is suffering from a lack of rain. Rivers, such as the Colorado, are at their lowest levels in years. Area lakes are suffering. Other parts of the country are experiencing flooding, with excessive rainfall.

Obstacles

If a person has always had easy access to nonpolluted water in sufficient quantity, it is difficult for them to realize that both water pollution and clean water in sufficient quantities are problems. Another obstacle is a belief that too many regulations to control these problems can cause job loss and stifle economic growth. In addition, many industries are opposed to new laws concerning water pollution. And new research is needed, providing up-to-date information on water concerns, especially how climate change will affect all aspects of water, from its sources to its many uses.

Applying Critical and Creative Thinking

Reflecting on the research we have done on issues related to water and the environment, we can conclude that three major problems must be dealt with:

1. Overuse of water.
2. Water pollution.

3. Water distribution, which will be dramatically influenced by climate change.

After reflecting on the history of water, the current issue of having adequate supplies of water in the future, as well as the effects of water pollution, we turn to identifying the beliefs and values that undergird the information we have gathered.

Beliefs and Values About Water

- We must recognize that water is essential to life.
- We must not take for granted access to fresh water.
- We must recognize the importance of preventing water pollution.
- We must recognize that climate change is real and will greatly affect the distribution of water.
- We must learn ways to conserve water.

Selecting and Applying Action Steps

After studying the information we have collected, and reflecting on the associated beliefs and values, we now consider possible action steps to the threefold problem. First, overuse of water. One way to address this problem is by limiting the *direct use* of water, which is the water that comes from the tap—water used for drinking, cooking, and washing dishes and clothes. Here are some ways to cut back on direct use of water:

- Use water-saving toilets.
- Install a water-saving shower head.
- Take shorter showers.
- Turn off the tap when while brushing teeth.
- Only do laundry when necessary.

Another way to address the overuse of water problem is by reducing the *indirect use of water*, which can include the following:

- Although a seemingly small action, eat fewer almonds. It might surprise some that it can take more than a gallon of water to grow one almond.[10]
- Eat less red meat (which uses large amounts of water to produce). Protein substitutes include beans and peas, for example.
- Eat chicken, which requires less water to produce than red meat.
- Drink water rather than bottled drinks such as colas, and avoid drinking bottled water. Bottled water is many times more expensive than tap water, and plastic bottles, when not recycled, contribute to environmental pollution.[11]
- Eat less processed food, as it requires water for each stage of its production.
- Eat more locally produced food—transporting food long distances uses large amounts of petroleum, which requires water for its production.
- Use natural items for cleaning such as vinegar and baking soda. They are phosphate free, and generally require less water when used.

As to the continuing problem of water pollution, people can support the Clean Water Act by knowing its provisions and reporting violators to appropriate authorities. As discussed, agriculture is a major polluter of rivers, streams, lakes, and groundwater because of fertilizer and pesticide runoff, as well as runoff from animal waste. Educational programs for farmers can help them understand the severity of the problem and help them take steps to prevent runoff. Educational programs, along with strict regulations, can help industrial-size animal agriculture operations prevent runoff, as well as prevent pollution of groundwater, which can make drinking water from wells dangerous to consumers.

Medium-size dairies should be emphasized. The group working with the Wisconsin Academy of Sciences, Arts and Letters studied the future of water in Wisconsin and concluded that "moderate-sized" agricultural operations, including those that practice rotational grazing are "less capital intensive and environmentally benign." This approach to agriculture is commonly called sustainable agriculture. [12] See chapter 5 to learn about this more friendly environmental approach to farming.

Looking at water distribution, the United States can roughly be divided into three sections in terms of annual rainfall. Western states such as Colorado receive on average 17 inches of annual rainfall, and Wyoming receives 13 inches. Midwestern states such as Wisconsin receive, on average, 34 inches of annual rainfall; Indiana receives 42 inches. Eastern states such as Vermont receive, on average, 43 inches, while New York receives 40 inches.

Because of the limited average rainfall, and with increasing agricultural irrigation along with a growing number of industrial-size dairy farms, western states have passed extensive water laws dealing with who has rights to water, and how much they can use. In Midwest states, such as Wisconsin, where industrial-size dairy farms as well as industrial-size crop production are on the increase, more extensive water laws likely need to be passed to assure that lakes and rivers do not dry up as a result of poorly regulated irrigation water use. The National Agricultural Law Center has done a good job of outlining the various water laws that now exist, the majority of them in the western states. [13]

Realizing that climate change will have a considerable effect on which parts of the planet will have little water and which will experience floods that devastate coastal cities and low-lying islands, we must consider what efforts we can take to help slow down climate change.

In summary, everyone needs to be concerned about the future for water, no matter whether you live in a major urban center or you are a farmer. Each of us has a responsibility to do some careful critical and creative thinking about water, and take the necessary action steps to make sure that water is protected, not polluted, and equitably distributed.

CHAPTER 8

Energy

Building Up Steam for Sustainability

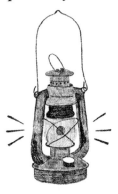

Background

As the world experiences the effects of climate change, a focus on energy sources becomes increasingly important. Energy—how we use it, where it is created, and how it is distributed—has become a major issue for every citizen to think about critically and creatively, consider alternative sources, and then take action steps.

In the mid-1930s, when I was a kid growing up on a farm, our major energy sources were a team of horses, a pile of stove wood, a gallon can of kerosene, the wind, and human power. Our team of draft horses did all the heavy work on the farm, from plowing and cultivating to harvesting the crops. We heated our farmhouse and the pump house with woodstoves, using wood we harvested each fall from the woodlot just north of the farmhouse. The gallon can of kerosene provided the fuel for two barn lanterns and three kerosene lamps that lighted the farm home. It lasted for a week. A windmill pumped our water from deep down in the aquifer. All was well as long as the wind blew. When the wind didn't blow, we were in trouble.

Much of the work around the farm was provided by the family members, and this ranged from walking behind a plow pulled by the team of horses, removing weeds with a garden hoe, chopping wood, and milking cows by hand as well as all the work associated with harvesting hay and grain. My mother's work included caring for the chicken flock and the vegetable garden as well as taking care of our health, washing and mending our clothes, and cooking our meals.

We hauled our grain to the mill in Wild Rose, Wisconsin; the mill was powered by waterpower. Early millers had first constructed a dam to back up the Pine River, forming a millpond that provided a dependable source of water to power the mill. In 1908, Edward Hoaglin, the miller, installed a generator, also water powered, that provided limited electricity for the village of Wild Rose. By limited, I mean that electricity was only available in the evening and during the noon hour, when the mill wasn't using its waterpower to grind grain.

It wasn't until 1947 that electrical power came to our farm, largely the result of President Franklin Delano Roosevelt signing Executive Order No. 7037 on May 11, 1935, which established the Rural Electrification Administration. In 1936, the Rural Electrification Act was passed by Congress, which created a lending program for rural electric co-ops. Unfortunately, World War II caused a slowdown because of a labor shortage, and supplies such as copper wire were in short supply. Thus, few farmers were able to get electricity to their farmhouses and barns until after the war.[1]

One of our neighbors had wind-powered electricity. He installed a Wincharger, which was a windmill that generated electricity that was stored in a room full of batteries. The Wincharger was invented in 1927; the first ones recharged 6-volt batteries used to power radios. By the late 1920s, many farmers had radios, including my folks. By 1937,

Winchargers were able to provide 6-volt, 12-volt, 32-volt, and 110-volt electricity. Our neighbor, Andrew Nelson, had a 32-volt system that provided electric lights and some limited electric motor use such as operating a washing machine.[2]

In the spring of 1947, I was in eighth grade at our community one-room country school. One afternoon, I came home and heard my dad say, "We are hooked up," meaning we had electricity. Electric lights came on in the Apps farmhouse, in the barn, granary, chicken house, and pump house. An electric motor powered the pump, and as long as electricity was available (storms sometimes interrupted service), we had water. It would be several more years before we had indoor plumbing in the farmhouse. Soon after we were "hooked up," an electric motor powered a milking machine, and we no longer milked cows by hand.

There was more to having electricity than making life a bit easier on the farm. Now, as I began high school in the fall of 1947, I didn't have to be ashamed of inviting friends from the village to come out to the farm. We had what many of them had already taken for granted—electricity. By the late 1930s, the village of Wild Rose became part of the grid with electricity available all the time, not just in the evenings and during the noon hour.

History of Energy Sources in the United States

Just as we used woodstoves in the 1920s, 1930s, and 1940s to heat our farmhouse and prevent the pump from freezing, wood was an important source of energy from the early settlement days continuing until the mid- to late 1800s. Water powered the early saw- and flour mills. By the late 1800s, coal became the dominant energy source. Petroleum became an important energy source by the mid-1900s, especially gasoline and diesel fuel, which powered the nation's cars, trucks, tractors, and locomotives. Natural gas consumption also rose rapidly during this time, especially for heating homes. By the mid-1900s, coal was used as a primary energy source for electric power generation. Nuclear electric power also emerged during this time.

Three major fossil fuels—petroleum, natural gas, and coal—provided up to 87 percent of the nation's energy in 2012.[3]

Hydraulic Fracturing: "Fracking"

One example of an energy source is known as fracking. Oil people have known about it for more than a hundred years—it's a process that cracks open rocks deep below the Earth's surface to access trapped fossil fuel deposits of natural gas or crude oil. The fracking process begins by drilling a long vertical well that can extend a mile or more into the Earth. As the well nears the rock formation containing natural gas or petroleum, the drilling gradually turns horizontal and may extend for several thousand more feet. At this point in the process, a fracking fluid is pumped in at a pressure great enough to create fractures or openings in the rock. The petroleum or natural gas then flows to the surface where it is gathered, processed, and transported. The contaminated wastewater is stored in pits or disposed of in underground wells. The fracking fluid consists of 97 percent water plus chemical additives and a special sand that keeps the fractures in the rock formation open after the pressure from the injection decreases.[4]

Proponents of fracking argue that natural gas is a bridge fuel toward getting to 100 percent clean energy, that fracking is a safe and affordable way of extracting it, and the process of fracking is allowing the United States to move toward energy independence.

On the negative side, fracking targets fossil fuels, including natural gas or petroleum. A growing number of people believe the country should move away from all fossil fuels, and put the kibosh on fracking. Besides, fracking can pollute groundwater, cause earthquakes, and increase greenhouse gases, which contributes to climate change.[5] A further negative of fracking is its need for a special sand. Mining this sand in Wisconsin and Minnesota has created several environmental challenges.

Sand Mining. Along with fracking comes sand mining. Once rocks containing natural gas or petroleum are fractured, the fracture must be "propped" open for fracking to be successful. To prop open a fracture, a special sand of the right size, shape, and strength is used. Western Wisconsin is the site for the perfect sand for fracking, along with several counties in southeastern Minnesota. Although it provides a short-term economic boost to the communities where it is found, the development of sand mines has itself become an environmental challenge. Sand mining is associated with both air and water pollution. Once the sand is mined, it is usually processed on-site, which involves washing and separating the sand into grain size suitable for hydraulic fracturing. The sand is then shipped to the states where fracturing mines are located.[6]

Some of the environmental challenges of sand mining include sending fine silica particulates into the air, which causes health problems; abandoned sand mining pits; polluted water left from washing the sand; and air pollution from heavy truck traffic.[7]

Mix of Energy Sources

In 2020, the United States used a mix of energy sources. Primary sources included fossil fuels, nuclear power, and renewable sources. Renewable energy sources included wind, solar, hydropower, geo-thermal, and biomass; alternative sources include ethanol, biodiesel, landfill gas and biogas from municipal solid waste, and wood and wood waste. In 2020, renewable energy production and consumption reached record highs.[8]

The Environmental Problem

Electricity is a clean and relatively safe form of energy. The environmental challenges from electricity relate to how it is generated and how it is moved via transmission lines from the power plant to the end users. In the United States, three major energy sources are used for the generation of electricity: fossil fuels, nuclear energy, and renewable energy sources such as wind, hydropower, solar power, and biomass.

Energy Sources for Electricity Generated in the United States, 2020

- Natural gas, 40 percent
- Nuclear power, 20 percent
- Renewables, 20 percent
- Coal, 19 percent

Of the renewables, 8.4 percent was wind, 7.3 percent hydropower, 2.3 percent solar, 1.4 percent biomass, and 0.5 percent geothermal.[9]

The Clean Air Act (1970) regulates the amount of air pollutant emissions a power plant creates. Administrated by the US EPA, the Clean Air Act has substantially reduced the number of air pollutants in the United States. But power plants that burn fossil fuels or materials made from fossil fuels continue to be a major source of air pollution in the United States.

Energy Users in the United States, 2020

- Industrial (33 percent)—manufacturing, agriculture, mining, and construction
- Transportation sector (26 percent)—cars, trucks, aircraft, ships, buses, and trains
- Residential (22 percent)—homes and apartments
- Commercial (18 percent)—offices, malls, stores, schools, hotels, hospitals, warehouses, restaurants, and places of worship[10]

Amount of Energy Used in Recent Years

In 2018, the United States recorded its highest use of energy; it was 1 percent lower in 2019, and 7 percent lower in 2020, mostly due to the COVID-19 pandemic. The transportation sector experienced the greatest decrease, about a 15 percent drop in 2020. Interestingly, the average amount of energy consumed per capita peaked in the late 1970s, as the US population increased, and total annual energy consumption trended upward over time. It was relatively flat from the late 1980s through the 2000s and has decreased each year since. Some of the reasons for the decrease in per capita energy use include:

- Population increases in the warmer part of the country contrasted with population decreases in colder regions, resulting in lower heating energy consumption.
- Fuel efficiency in vehicles resulting from the Corporate Average Fuel Economy standards.
- Increases in the energy efficiency of household appliances, electrical equipment, building insulation, and improved building energy codes.[11]

Obstacles

One of the important obstacles in combating the environmental impact of energy production, transmission, and use is attitude. If my electricity is on each day, I don't worry about it. Or, the old saw, if it's not broken, why fix it? Another obstacle is an overall lack of understanding concerning how energy is produced and distributed, and the relative challenges each energy source places on the environment.

Beliefs and Values About Energy

Here are some beliefs and values that can prevent people from taking an active role in combating the impact of energy on the environment:

- Fossil fuels, such as oil and natural gas, are dependable sources of energy.

- Technology, such as fracking, has created new sources of fossil fuel energy and should be applauded and encouraged.
- Ethanol added to gasoline makes it more environmentally friendly and will allow gasoline to be used as the primary energy source for the travel industry for years to come.
- Most alternative energy sources are a fad, and in the long run, will not be economically feasible.
- Other issues, such as jobs, national security, and the economy, are far more important than worrying about the environmental impact of energy sources and their uses.

Applying Critical and Creative Thinking

After examining the history of energy and its sources over time, and the current situation involving energy and the environment, we might ask the following questions: What are the specific impacts on the environment from today's energy sources? What are alternative ways of lowering these impacts? What are the environmental impacts of how energy is used and the amount that is used? What are alternative ways of cutting the amount of energy used?

Here are some beliefs and values that might be identified. You likely have others to add to the list.

- Individuals, through their reduced use of energy, can have a positive impact on the environment.
- Everyone should be concerned about the sources of energy, as well as the amount used and the potential impact on the environment.
- For the future of the planet, it is important for each of us to consider alternative sources of energy—perhaps some that have not been identified.

- We should not hesitate to leave behind some traditional sources of energy, especially those that are tied to fossil sources.
- We must all realize that the many dimensions of energy are as much an environmental concern as an economic one.

After reflecting on the alternative beliefs and values above, and deciding something should be done to make energy sources and uses less impactful on the environment, the next question is: What are some alternatives to the traditional ways of doing things? Let's look at renewable sources of energy.

Renewable Energy

Renewable energy can replace traditional energy, which mainly relies on fossil fuels, and is much less harmful to the environment. Renewable energy sources include:

- **Wind.** Wind is captured by using turbines and converting its energy into electricity. It is a clean energy source—it doesn't pollute the air or water as do some other forms of energy. Wind energy doesn't produce carbon dioxide or release any other byproducts into the atmosphere. One of wind energy's main negatives: wind farms are often located in remote areas and require transmission lines to move the energy to where it is needed. Some citizens oppose them because the wind turbines can dominate a skyline and generate some noise and shadows. They also can threaten birds and bats, which are sometimes killed by the turbine blades. And even though they are generally located where the wind regularly blows, there are times when the wind doesn't blow and thus no electricity is generated.
- **Solar.** Solar panels capture the sun's energy and convert it into heat, electricity, or hot water. An obvious advantage to solar energy—there is a limitless supply of it. A disadvantage—the

upfront cost for solar panels can be considerable. Also, home-owners who want to switch to solar energy need to have a place on their homes where the panels can face the sun.

- **Hydropower.** One of the oldest forms of renewable energy, hydropower is created when a river or stream is dammed, and water flows through the dam turbines to produce electricity, or it can produce energy directly; for example, when water is used to power a mill. Hydropower was one of the first energy sources to produce electricity. A disadvantage of hydropower is its dependence on water levels. Seasonal variations in precipitation and droughts, especially long-term droughts, can have detrimental effects on hydropower production. Another disadvantage is the disruption of water systems, which with changing water levels, can have a negative effect on fish and other wildlife.[12]

- **Biomass.** This renewable energy source comes from recently living plants and organisms. Energy can be created by burning the biomass, or by capturing methane gas produced by landfills and large animal dairy sites, which employ methane converters to convert animal manure into methane. It can also be produced by burning wood.

- **Geothermal.** This energy source comes from heat that is trapped beneath the Earth's crust. It escapes through volcanoes, hot springs, and geysers. Geothermal plants, most of them located in the western United States, produce geothermal energy by using steam from the heated water that rises to the Earth's surface and is used to operate a turbine. Geothermal energy is not as common as other renewable energy sources. Major disadvantages include the cost of building the infrastructure to the capture the energy, and the possibility of earthquakes that would destroy the system.[13]

- **Ethanol.** In the United States, ethanol, which is a type of alcohol, is mainly produced from corn. In 2020, 40 percent of the US corn crop was used for ethanol production.[14] Ethanol can also be produced from waste materials such as wheat straw, sugarcane, and wood residues. Today, nearly all US gasoline sold is blended with ethanol. A major advantage of ethanol is that it reduces the amount of greenhouse gas emissions from cars and is thus a cleaner fuel than gasoline and diesel fuel produced from crude oil. One disadvantage is the need for engine modifications for using a mixture of ethanol that exceeds 10 to 15 percent. Small engines, such as those used on lawn mowers, rototillers, and chainsaws, could be ruined with even a low mixture of ethanol with gasoline. A major disadvantage of ethanol made from corn is the vast acreages of corn grown for ethanol that would otherwise be grown as a food source. Although the ethanol itself can be viewed as environmentally friendly, growing corn is not. Growing corn requires commercial fertilization, pesticide control, and oftentimes, irrigation, all of which are viewed as environmentally unfriendly.[15]

Selecting and Applying Action Steps

Having researched the history of energy, learned about past energy sources and how energy was used, examined sources and uses today, and considered the obstacles to combating environmental impacts, what action steps can be taken to lessen energy's impact on the environment? Research is quite clear that the main source of energy for many years has been fossil fuels (petroleum, natural gas, and coal), and they hurt the environment. These energy sources must be left behind if the country wants to cut back on greenhouse gas emissions that contribute to climate change. What are some action steps that individuals who have critically and creatively thought about the problem can take?

- **Switch to cleaner energy sources.** As an example, many home-owners can benefit from solar panels on their homes for self-sufficient electrical generation. More options are also becoming available for sources such as small wind systems and "microhydropower"—especially if you're building a new home.[16]

- **Become more energy efficient.** There are simple things you can do such as using LED light bulbs. Only 10 to 20 percent of the energy incandescent bulbs use is converted to light; the rest creates heat. Eighty to 90 percent of the energy LED bulbs use is converted to light. Also, LED bulbs consume 80 to 90 percent less energy than incandescent lights.[17] You can also install a low-flush toilet and a low-flow showerhead. Other steps you can take include making sure your furnace is in good running condition, installing a clothesline, and running your larger appliances at night instead of during the day.[18]

- **Be an informed consumer.** Purchase energy-efficient home water heaters, refrigerators, televisions, stoves, washers, and air conditioners.

- **Make improvements.** Have a professional energy auditor check your home and identify areas that need improvements to increase energy efficiency. In addition, there are often government or utility company–sponsored programs that will assist people in making their homes more energy efficient. For example, for low-income customers, there is a government program called the Weatherization Assistance Program that will use state-of-the-art technology to make a dwelling more energy efficient.[19] Check with your local government or utility provider for other programs.

- **Consider transportation changes.** Consider driving an electric vehicle rather than one that runs on fossil fuel. Use public transit when available. Carpool, ride a bike, or walk.

- **Get involved.** Support lawmakers and other policy makers as they promote environmentally friendly energy sources and energy distribution systems.
- **Support sensible regulation.** Rules and regulations must be put in place to assure that commercial buildings, as well as all public buildings, meet clean-air as well as energy-efficient requirements. Policies should be put in place to encourage that all transportation systems—rail, air, highway (both commercial and private)—meet energy-efficient as well as clean-energy requirements.

Of course, individuals interested in reducing energy use can insist that the energy produced be "clean" energy—meaning it is environmentally friendly. They can do many things at home to cut down on energy use, and their voices can also help influence how industries, including agriculture and manufacturing, use energy. And those voices can influence what the sources of energy ought to be, focusing critical attention on moving beyond fossil-fuel sources to renewable energy sources.

CHAPTER 9

Air Quality

*Finding It's Too Soon
to Breathe a Sigh of Relief*

Background

I'll confess. In the 1960s and 1970s, we burned our trash. In fact, those fifty-five-gallon backyard burn barrels we used have become artifacts of country life at that time. In my defense, we burned before we had rolling trash carts that we could take to the curb and garbage trucks to then haul away our properly sorted waste and recyclables.

Without those modern waste-jettisoning amenities, we burned plastic, wood, sensitive documents, and all sorts of household waste. If it burned and fit, it went into the barrel.

While we were careful to secure the required local burning permit, others were much sneakier, avoiding the permit fee and hoping no one would notice the noxious smell and smoke curling up from behind their barn. Today, some people still resort to backyard burning to avoid trips to landfills—and even for more nefarious purposes such as disposing of evidence of a crime. Burning leaves remains a common way to dispose of leaf litter—a better option would be to mulch the leaves and return organic matter to the yard.

At the time when we were burning, we weren't thinking about the chemicals, such as formaldehyde, hydrogen cyanide, and carbon monoxide, that we were releasing into the air. We didn't know that burning waste could release harmful chemicals like dioxins—a proven human carcinogen. Burning barrels, after all, are inefficient incinerators in large part because they burn at much lower temperatures than commercial incinerators and are open to the environment, allowing for greater particle pollution. Commercial incinerators are also regulated by the EPA in the United States with the requirement to have some level of pollution control systems in place.

On the farm, dioxins were of special concern. When dioxins become airborne, they can mix with rain and then be taken up by crops that farm animals eat. Then, humans can become exposed to dioxins when they eat the fats found in animal products. Dioxins also enter surface waters through rain.

A study by the EPA found that the amount of cancer-causing dioxin and furan emissions from fifteen households burning trash each day is the same as emissions from a 200-ton-per-day municipal waste incinerator that uses high-efficiency emission control technology.[1] Today, most outdoor waste burning is banned. In fact, we retired our barrel decades ago.

Instead, when we think of air pollution, we think of the black smoke emitting from coal plants. We think about exhaust belching out of the pipes of older vehicles and semis. And in recent years, with drought and climate impacts, we think of wildfires. These toxic emissions can affect local residents as well as those fighting the climate impacts such as firefighters responding to wildfires and EMTs, nurses, and doctors who treat those impacted by smoke exposure and who have sustained fire-related injuries. But the effects of air quality can also be felt thousands of miles away from a fire.

Good air quality pertains to the degree to which the air is clear and free from pollutants such as smoke, dust, and smog, among other impurities. Poor air quality can harm human and environmental health.

Although many local governments had passed air pollution legislation already, in 1955, the federal government decided that the problem needed to be addressed on a national level. Air, after all, doesn't recognize the boundaries of state lines. As a result, Congress passed the Air Pollution Control Act of 1955. This represented the first piece of federal legislation on this issue in the United States. The bill called out air pollution as a national problem and requested research and action to address it.

The Clean Air Act of 1963 was the first federal legislation to address air pollution control. It was implemented by the US Public Health Service and called for research into monitoring for and controlling air pollution.

In 1967, the Air Quality Act was passed, expanding federal authority for enforcement in areas subject to interstate air pollution transport. The federal government also launched ambient monitoring studies and began stationary source inspections.

By 1970, air pollution concerns were front and center in people's minds, as seeing is believing. Many of the nation's cities faced dense smog and acid rain, prompting Congress to react to the crisis and pass legislation buoyed by the national environmental movement that was underway.

The Clean Air Act of 1970 led to federal and state regulations to limit emissions from stationary and mobile sources. "The US EPA was created on December 2, 1970, to implement the various requirements included in the Clean Air Act." The Clean Air Act was amended in 1977 and 1990.[2]

The Montreal Protocol on Substances That Deplete the Ozone Layer was created in 1987 to address the fact that "good" ozone was being eroded by industrial chemicals. The problem was highlighted in the media when "a gigantic hole appeared in the ozone layer over Antarctica in the mid-1980s, allowing dangerous levels of UV radiation to reach the surface."[3] The Montreal Protocol gradually phased out the production and use of CFCs and HCFCs (hydrochlorofluorocarbons).

"The Earth's ozone layer would have collapsed by 2050 with catastrophic consequences without it," National Geographic reported on the Protocol's 30th anniversary. "There would have been an additional 280 million cases of skin cancer."[4]

The Environmental Problem

According to NOAA, "Poor air quality is responsible for more than 100,000 premature deaths in the United States each year and costs from air pollution-related illness are estimated at $150 billion per year." Outdoor air pollution and particulate matter have been classified as carcinogenic to humans by the International Agency for Research on Cancer.[5]

The EPA has a grave list of ailments poor outdoor air quality can cause:

- Aggravates asthma
- Increases respiratory issues like coughing or irritation of the airways
- Decreases lung function
- Causes irregular heartbeat
- Damages the liver, kidney, and central nervous system
- Leads to premature death in people with lung or heart disease[6]

The EPA developed the Air Quality Index (AQI) as a way to visually share information about the health effects of five common air pollutants and how to avoid those effects. "Action days" are usually called when the AQI rises into unhealthy ranges. Sometimes air quality can be so bad that, for some people, it may be dangerous to engage in usually healthful outdoor exercises such as walking, running, or playing golf. For those at risk from air pollution—children, the elderly, and those with underlying health conditions—it is necessary to check the local air quality before heading out to exercise.

Research by the University of Chicago's Energy Policy Institute found that "air pollution shaves off 2.2 years of average life expectancy worldwide," and those living in "the most polluted areas could see their lives cut short by five years or more." Driving the news of life expectancy impact is *The Air Quality Life Index*, published in September 2021, which

> shows that the burden of harmful air pollution is unevenly distributed—with China making rapid, measurable progress in cleaning up its air and other global hotspots now emerging in South Asia and Sub-Saharan Africa . . . . Working unseen inside the human body, particulate pollution has a more devastating impact on life expectancy than communicable diseases like tuberculosis and HIV/AIDS, behavioral killers like cigarette smoking, and even war.[7]

Types of Air Pollution

Particulate Matter

Particulate matter (PM) refers to airborne particles such as dust and smoke. Some of these particles are the result of vehicle use, industrial operations, construction sites, and burning wood. Other particles are formed when sunlight and water vapor react with gases.

This pollution can cause serious health effects such as chronic bronchitis, asthma, coughing, and even heart attacks. In addition, air pollution has serious environmental impacts such as making lakes and streams more acidic. It also can strip nutrients that plants need from soil.

The two subcategories of PM you may have heard of are PM10 and PM2.5. The size of the matter is what makes them different. The numbers to the right of the "PM"—10 and 2.5—indicate the aerodynamic diameter of the particles. PM10 is often called coarse dust and PM2.5 is called fine dust. Unlike PM10, PM2.5 not only enters your lungs, but can enter the bloodstream.

Then, there is PM0.1, or ultrafine dust, which is less frequently talked about as there is little data available. While less is known about it, it is widely believed that given its size, PM0.1 may cause even more health problems than PM2.5 and PM10.

Ozone

The second most common air pollutant in the United States is ozone, and there are "good" ozone and "bad" ozone.

Good ozone lives in the upper atmosphere that protects the planet from the sun's ultraviolet rays. Bad ozone is found closer to the planet's surface and comes mostly from human activities such as burning fossil fuels. It is formed by pollutants emitted by vehicles, power plants, and other industry. When these airborne pollutants encounter sunlight and warm temperatures, they react to become ozone.

Even at low concentrations, ozone can cause respiratory illnesses. Scientists also are increasingly understanding how ozone damages plants and trees and harms everything from soil microbes to wildlife.

Ozone damage can alter the timing in the season when leaves fall. In the United States, black cherry, quaking aspen, and white pine are among the tree species most negatively affected by ozone.[8]

Ground-level ozone is a greenhouse gas. It's the "third worst [greenhouse gas] after carbon dioxide and methane," reports Yale's School of the Environment. At the "ground level," ozone becomes "the most damaging pollutant in the world," environmental Professor Evgenios Agathokleous told Yale's School of the Environment. "It induces the most widespread damage to plants, and it's a very serious threat to biodiversity."[9]

We might think of ozone as a city-specific problem. But that's not the case. National parks and other wild areas are suffering too. "Ozone levels at Sequoia [National Park] and the adjacent national park, King's

Canyon, are among the highest in the United States, thanks to smog that blows in from the urban areas and farming and industrial activity in the San Joaquin Valley below. Smog levels here are sometimes as high, or higher than they are in Los Angeles."[10]

High ozone levels are also expected to lead to declining global food production. One recent study offers that "impacts of ozone air pollution and temperature extremes on crop yields [include]: Spatial variability, adaptation and implications for future food security." It goes on to predict "that by 2050 wheat production will decline by 13 percent, soybeans by 28 percent, and corn by 43 percent because of rising temperatures and ozone."[11]

Studies of forests and related ozone levels in California's San Bernadino Mountains have shown that air pollution increases forest susceptibility to wildfire because "ozone-sensitive, fire-resistant species of pine were replaced by species that were more likely to burn."[12]

In the first 10 months of 2021, the National Interagency Fire Center reported 48,366 wildfires across the country that burned more than 6.5 million acres.[13] As many as 90 percent of wildland fires in the United States are caused by people, according to the US Department of Interior. The US Forest Service cites major sources of human-caused fires such as campfires left unattended or thought to have gone out, debris burning, downed power lines, discarded cigarettes, and even arson. The remaining 10 percent are started by natural causes such as lightning strikes or lava.[14]

Worsening air pollution from climate change is sometimes referred to as the "climate penalty."[15]

Acid Rain

Acid rain, or acid deposition, includes "any form of precipitation with acidic components, such as sulfuric or nitric acid that fall to the ground from the atmosphere in wet or dry forms—rain, snow, fog, hail, or even dust."[16] Acid rain is formed when rainfall or atmospheric moisture mixes with elements and gases that cause the moisture to become more acidic than normal.

"Pure water has a pH of 7, and generally, rainfall is somewhat on the acidic side (a bit less than 6). But, acid rain can have a pH of about 5.0–5.5, and can even be in the 4 range in the northeastern United States, where there are a lot of industries and cars."[17]

Acid rain impacts plants, and seeds do not grow well in more acidic soils. Acid rain alters soil pH, strips vital nutrients from the soil, and eats away the protective waxy cover of leaves, making them more likely to be exposed to pests and drought. Soil pH testers are an important tool for many gardeners. Many plants and trees are damaged by acid rain. Some fish and amphibians are unable to reproduce in an acidic environment. In addition, acid rain accelerates building weathering and damages stone-work and cultural artifacts.

Seventy-five years ago, acid rain was a problem that largely affected the eastern United States. "It began in the 1950s when Midwest coal plants spewed sulfur dioxide and nitrogen oxides into the air. . . . As acid rain fell, it affected everything it touched." It poisoned lakes, and low pH waters killed fish and nearly wiped out fish-eating birds such as loons.[18]

Obstacles

- Air pollution took a long time to regulate because we did not have enough data to know its full impacts, including impacts on human health, plants, and animals. With more research on a wide scale, we know much more. We are now much better equipped in that regard, but some areas still have too little data to know the extent of the air pollution problem there.

- Most countries have some form of air-quality regulation, but air pollutants from one country can drift to neighboring countries and from one state to another.

- Air pollution is a byproduct of many processes necessary to our economy and our lives, such as transportation and electricity.

- Reducing air pollution comes at a cost. (But arguably, the value of Clean Air Act health benefits far exceeds the costs of reducing pollution.)

Reducing Air Pollution

"Reducing air pollution improves crop and timber yields, a benefit worth an estimated $5.5 billion to those industries' welfare in 2010," according to the peer-reviewed March 2011 EPA study.[19] Cleaner air means fewer air pollution–related illnesses, which in turn means less money spent on medical treatments as well as lower work and school absenteeism. "Better visibility conditions in 2010 from improved air quality in selected national parks and metropolitan areas had an estimated value of $34 billion."[20]

Applying Critical and Creative Thinking

Air pollution's toll on human health across the world is well-documented. It also has economic and social implications. Now that we have established the widespread critical impacts of air pollution, and provided some examples of legislation and technology solutions addressing it, we might ask the following questions: What can we do to ensure we are basing our thinking and actions on the most current data? What technologies will be needed moving forward to continue to address air pollution? What can we do for communities that, while largely impacted by air pollution, lack the resources—such as funding and access to sophisticated technologies—to address the problems?

Thanks to intensive monitoring, we now know, for example, that air quality has improved in many areas since the passage of the Clean Air Act and Montreal Protocol. In response to increasingly strict environmental regulations beginning in the 1970s, gasoline- and diesel-powered vehicles were equipped with catalytic converters that transform more dangerous air pollutants into less harmful pollutants.

Monitoring proves that "Clean Air Act programs have lowered levels of six common pollutants—particles, ozone, lead, carbon monoxide, nitrogen dioxide, and sulfur dioxide—as well as numerous other toxic pollutants. Between 1970 and 2020, the combined emissions of the six common pollutants dropped by 78 percent."[21]

Airborne lead pollution, which was a serious health concern before the EPA required phasing out of lead in gasoline, now meets national air quality standards in most areas of the United States. Continuous and more advanced monitoring, such as that done by using satellites, provides data more quickly today and helps us better understand the complex nature of air pollution. We can apply this data from decades of air quality monitoring to our critical and creative thinking skills to continue to explore clean energy technologies.

Reduction of Vehicle Emissions

"Compared to 1970 vehicles, new cars, SUVs and pickup trucks are roughly 99 percent cleaner for common pollutants (hydrocarbons, carbon monoxide, nitrogen oxides, and particle emissions), while Annual Vehicle Miles Traveled has dramatically increased."[22] And "starting in the 2014 model year," for example, "locomotives are 90 percent cleaner than pre-regulation locomotives."[23]

The Clean Air Act has led to clean technologies and innovation in reducing emissions and controlling the costs of doing that work. Catalysts, scrubbers, and low-VOC paints and coatings are examples of products that are widely used today. "Stationary sources now emit about 1.5 million tons less toxic air pollution per year than in 1990."[24]

But despite some gains, there is no reason to breathe a deep sigh of relief just yet. There are still many areas throughout the world, including the United States, where populations face unhealthy levels of air pollutants. There is unfinished business.

In July 2021, the EPA published a free and open-source web resource featuring interactive maps and support materials that combine information on air pollution emitted by fossil fuel–fired power plants with demographical data.[25] This provides information to the public, policy makers, and businesses.

The Power Plants and Neighboring Communities initiative educates the public and political leaders so that they can better understand the impacts of air pollution in struggling communities. "We know air pollution affects some people worse than others. Achieving environmental justice starts with improving our understanding of the impacts of air pollution, especially in overburdened and historically underserved communities," said EPA administrator Michael S. Regan when announcing the interactive website. "This web resource equips users with actionable, science-based data on air quality in communities near power plants, many of whom are suffering the worst from pollution." The groups tracked include low-income people, people of color, those with less than a high school education, those facing language barriers, children under age five, and those over sixty-four.[26]

Selecting and
Applying Action Steps

- Keep your vehicle tuned up, use mass transit, bike or walk, and switch to a more fuel-efficient vehicle such as a hybrid or all-electric vehicle. Establish a neighborhood carpool or ride share service.

- Use energy-efficient LED bulbs, and turn off lights and electrical equipment when you're not using them. Insulate your home. Use a programmable thermostat.

- Switch to solar and wind power. If you own a home, consider adding solar collectors to your roof. Check to see if your utility company offers incentive programs that let you purchase power from clean energy sources.

- Grow local. Food production contributes significantly to air pollution through transportation, for example. In turn, air pollution can impact food production.

- Be a conscientious consumer. For example, avoid products that contain palm oil. Forest fires are commonly set to clear land for the palm oil industry. Specifically, palm oil at "66 million tons annually . . . is the most commonly produced vegetable oil. Its low world market price and properties that lend themselves to processed foods have led the food industry to use it in half of all supermarket products. Palm oil can be found in frozen pizzas, biscuits, and margarine, as well as body creams, soaps, makeup, candles and detergents."[27]

- Use environmentally friendly landscaping tools such as an electric lawnmower. If possible, purchase bicycles for your transportation. Use environmentally safe cleaning products and paints. Combine errands and reduce the number of trips you take in your car. Reduce wood-burning fireplace and woodstove use when possible.[28]

- Be informed and inform others. The Clean Air Act and Montreal Protocol have shown us that it's possible to lessen air pollution by making changes in our personal lives and in our communities.

"Increasingly, our leaders must deal with dangers that threaten the entire world, where an understanding of those dangers and the possible solutions depends on a good grasp of science," wrote Isaac Asimov, an American author, biochemist, and prolific writer of science fiction and science books.

The ozone layer, the greenhouse effect, acid rain, questions of diet and heredity. All require scientific literacy. Can Americans choose the proper leaders and support the proper programs if they themselves are scientifically illiterate? The whole premise of democracy is that it is safe to leave important questions to the court of public opinion—but is it safe to leave them to the court of public ignorance?[29]

CHAPTER 10

Natural Resource Issues

Where Enjoying Meets Exploiting

Background

In the mid-1980s, a war came to northern Wisconsin. Dubbed the "Walleye War," battle was waged at boat landings over natural resources, and one in particular—walleye, a prized trophy fish that is the centerpiece of northern Wisconsin supper club fish fries.

For generations, the Ojibwe bands of northern Wisconsin have spearfished spawning walleye in the springtime. Because of treaties signed in 1837, 1942, and 1854 with the federal government, the bands have held the right to hunt, fish, and gather on lands in the northern third of Wisconsin.

A federal appeals court upheld these rights in 1983 (1983 *Lac Courte Oreilles v. Wisconsin [Voigt]* decision), and in 1988, a US District Court ruled that the Ojibwe had the right to off-reservation during the walleye spawning season.

As a result of these decisions, there was conflict between the Ojibwe and some non-Native anglers, who argued that it was unfair that bands were allowed to spear when walleye are spawning.[1]

Protesters to Ojibwe spearing flocked to the boat landings of those lakes being spearfished. Assassination threats were made against tribal members. Protesters held signs reading, "Spear an Indian, Save a Walleye." Racism was rampant.

In response, people held peaceful marches and ceremonies to support the Native fishers and decry the racism. I was in college when I joined the Witness for Non-Violence for Treaty and Rural Rights in Northern Wisconsin group and was trained to be a peaceful observer and to go to the boat landings to record through photos, writing, and video what I saw. It opened my eyes. Sometimes to believe something, you have to see it or hear it for yourself.

Natural resources are those that, as the name implies, come from nature—water, wood, air, land, plants, and animals. Some natural resources, such as plants that might seed and grow in a season, can be replaced quickly after they are used. Others, such as a tree that can take years to grow to a harvestable size for lumber, can take a long time to replace.

Renewable natural resources refer to "inexhaustible" resources such as solar and wind energy, biomass energy, and hydropower.

Other resources, such as fossil fuels, are nonrenewable, or "exhaustible," meaning that they cannot be replaced in our lifetimes, if at all. It takes millions of years for a dead organism, for example, to get converted into fuels such as oil, coal, and petroleum. Conservation is the practice of caring for natural resources.

The Public Trust Doctrine
Offers Public Rights Protection

For hundreds, if not thousands of years, the common law has recognized that some resources are so important they cannot be owned by anyone in particular—instead, they must belong to everyone together. Today, that law is called the "public trust doctrine."

Most versions of the doctrine require the state to manage resource commons for the benefit of the public. Throughout the centuries, the doctrine has evolved to recognize a responsibility to protect them for present and future generations.

In recent decades, the public trust doctrine has evolved substantially to address a greater variety of natural resources—from air to land and the waters—and a broader scope of public values, including navigational waters and ecological values.[2]

The Environmental Problem

The Earth's supply of natural resources is limited. As the human population increases, we are using up more and more of these natural resources. For example, extracting or mining for minerals is not easy and is made even more difficult the scarcer they become. Many mining methods devastate the environment. They destroy soil, plants, and animal habitats, and can pollute water and air when toxic chemicals used in the mining processes leak into the ecosystem.

When a mining company enters a community to extract raw materials, they often tout the advantages of the operation and minimize the potential negative environmental impacts in order to gain acceptance from the local community and government. Companies argue that they bring economic development and with that comes jobs and services such as an investment in the local tax base and schools. What happens when the mine is used up and the company leaves? The jobs vanish as well. There have been short-term economic gain with long-term negative environmental impacts. In place of jobs there may now be air pollution, soil degradation, and drinking water contamination.

Moreover, in some countries, profits from raw material production are used to finance armed conflicts. According to United Nations statistics, "In the last 60 years, at least 40 per cent of all intrastate conflicts have a link to natural resources, and that this link doubles the risk of a conflict relapse in the first five years."[3]

Other underground resources at risk of depletion include fossil fuels, such as coal, oil, and natural gas, which were formed millions of years ago. Without a shift to sustainable, renewable energy supplies, at the current rate of consumption, experts predict that we will run out of these resources in decades—not centuries. According to a prediction made in a 2019 publication from the Millennium Alliance for Humanity and the Biosphere at Stanford University, "The world's oil reserves will run out by 2052, natural gas by 2060 and coal by 2090."[4]

Above ground, other natural resources, such as timber, can be depleted when they fall victim to extreme harvesting methods such as clear-cutting. Overharvesting forests can lead to loss of biodiversity and habitat for birds and other animals, increase soil erosion, and removal of the carbon sink that trees provide.

The oceans are not exempt from exploitation, either. For example, overfishing, removing a species faster than it can naturally replace its population, is depleting our oceans of fish and sea creatures. Some fisheries are on the verge of collapsing as a result of large commercial fishing operations that have cast their nets wider and deeper in search of increased profits. Trawling—pulling nets with boats—leads to capturing critters not even intended for human consumption.

Obstacles

- Economic growth. While not all economic growth is harmful to the environment and leads to exploitation of natural resources, in some cases that does occur. According to the World Bank, "since 1995, the world economy has grown by 2.8% p.a. on average and international trade has increased sixfold, while the use of natural resources has come to a point where many renewable natural resources are classified as over-used."[5]

- More sophisticated technology allows natural resources to be extracted more quickly and efficiently than ever before. For example, it used to take long hours to cut down one tree using only axes and saws. Automated clearing methods have led to increased deforestation rates.

- Even some renewable resources require extensive energy, materials, chemicals, and in some cases, water, to extract from the Earth.

- The human population is increasing, putting greater pressure on natural resources. According to the United Nations, the world population was 7.6 billion in 2017. "This number is expected to rise to about 10 billion in 2050."[6]

- Cultures of consumerism. Materialism leads to a demand for the mining of gold and diamonds, for example.

- There is often a lack of awareness of ways to reduce depletion and exploitation of materials.

- Social inequities are created due to distribution of natural resources, including who has access to water and to clean water.

Applying Critical and Creative Thinking

Chief and Faithkeeper Oren Lyons (Onondaga, Seneca) writes: "We are looking ahead, as is one of the first mandates given us as chiefs, to make sure and to make every decision that we make relate to the welfare and well-being of the seventh generation to come. . . . What about the seventh generation? Where are you taking them? What will they have?"[7]

For the sake of the seventh generation—and even for the current generations of people, plants, and animals—we need to champion stewardship of our natural resources. The question is: Will you be a champion? And if so, that requires you to think critically and creatively about the issue.

In Wisconsin, home of the Walleye War, there is a shining example of bipartisan support to protect natural resources, provide public access to these resources, and champion stewardship.

The Knowles-Nelson Stewardship Program was created in 1989 to "preserve valuable wildlife habitat, protect water quality and fisheries, and expand opportunities for outdoor recreation." The program gives the Wisconsin DNR authority to purchase land and easement additions to state properties. The fund had bipartisan support when passed and is named after two Wisconsin conservation leaders: Warren Knowles (governor, 1965–1971) and Gaylord Nelson (governor, 1959–1963; US Senator, 1964–1981; and creator of Earth Day). "Today, stewardship dollars have been invested in every one of Wisconsin's seventy-two counties."[8]

Selecting and Applying Action Steps

- Help conserve fossil fuels so they are not depleted, but also because burning them pollutes the air.
- Subscribe to using renewable resources. Scientists are continuing to explore alternatives to fossil fuels. They have created ways to produce renewable biofuels and create power sources from the sun, wind, water, and geothermal sources.
- Learn about the roles and levels of government and whom to contact regarding natural resources management.

- Make interactions with policy makers a norm and encourage others to do it.
- Learn to use skills such as persuasive techniques. Learn how to frame a question—one at a time—so that it is clear and concise. Hone your listening skills and practice them. These skills will help you be more successful in your interactions with decision makers who are often pressed for time and flooded with different viewpoints.
- Know where to find reputable information on the status of natural resources in your area.
- Network with others who also want to make a difference in natural resources conservation.
- Be a steward. Remember and embrace Chief Lyon's Seventh Generation philosophy: one must be mindful of the effects of any decision or action on the future of the people through the seventh generation.

CHAPTER 11

Land Use

Seeing the Need for Resilient Road Mapping

Background

Long before shopping malls, skyscrapers, rows of identical-looking condominiums, and concrete highways covered the country, Indigenous Nations viewed land as something humans are a part of, rather than something that is separate.

"Native American teachings describe the relations all around—animals, fish, trees, and rocks—as our brothers, sisters, uncles, and grandpas. . . . These relations are honored in ceremony, song, story, and life that keep relations close," writes Winona LaDuke, Anishinaabe writer, economist from the White Earth reservation in Minnesota, and executive director of Honor the Earth, a national Native advocacy and environmental organization.[1]

And when we approach land use from that viewpoint—land is a family member (Mother Earth) rather than a commodity—we start to think of it as a relationship that carries responsibilities. "The truth is we cannot live apart from nature, we are *part* of nature."[2]

In "Indigenous Americans: Spirituality and Ecos," author Jack D. Forbes suggests that the land is a living thing. "Of course, the indigenous tendency to view the earth and other nonorganic entities as being part of *bios* (life, living) is seen by many post-1500 Europeans as simply romantic or nonsensical," he writes. "When Native students enroll in many biology or chemistry classes today they are often confronted by professors who are absolutely certain that rocks are not alive. But in reality, these professors are themselves products of an idea system of materialism and mechanism that is both relatively modern and indefensible."[3]

Aldo Leopold, considered the father of wildlife ecology, believed that the relationships between people and land are intertwined. In his book, *A Sand County Almanac*, Leopold wrote: "When we see land as a community to which we belong, we may begin to use it with love and respect. . . . That land is a community is a basic concept of ecology, but that land is to be loved and respected is an extension of ethics."[4]

Land as a Teacher

If we are paying attention, there are many lessons to learn from the land. In *Whispers and Shadows*, Jerry writes that the more people know about something, the more likely they are to prize it, protect it, and encourage others to do so as well. He says, "The more deeply we come to know our natural surroundings, the more willing we are to stand up and be heard when a wetland is threatened by development, an endangered species' habitat faces destruction, or mature trees are in danger of being cut down because they are 'in the way' of a new road."[5]

The land teaches us many lessons about the cycles of life, about the beauty of soil, and the importance of pride of place.

Every year, our prairie restoration project at our family farm, Roshara, teaches us to dig out our guidebook to identify a new grass or wildflower. We look for the effects of climate change on the land. We study imprints in the soil around our ponds for signs of wildlife. The lesson plan of the land includes learning by doing and not being afraid to get dirty doing it, embracing the mystery, and getting out for a walk.

The Environmental Problem

Regulations regarding land use were obviously foreign to Indigenous people, and there were few regulations among rural white European Americans. But when people started moving from the country into the cities, public land regulation became important. New York City passed the country's first zoning ordinance in 1916 and by the 1930s, most states had zoning laws.

Today, land use includes agriculture, housing, business and commercial, and recreation uses. The United States is a jigsaw puzzle of cities, farms, and forests. This 1.9-billion-acre puzzle includes public and private lands that often represent very different uses.

The USDA classifies six major types of land use in the United States. According to a USDA report, "Grassland pasture and range uses accounted for the largest share of the nation's land base, surpassing land in forest uses, for the first time since 1959." Although the shares of land in different uses have fluctuated to some degree over time, land area in the top three categories has remained relatively stable:

1. Grassland pasture and range
2. Forest
3. Cropland

The report notes that

52 percent of the 2012 U.S. land base (including Alaska and Hawaii) is used for agricultural purposes, including cropping, grazing, and farmsteads/farm roads. . . . Land dedicated to special uses such as state parks and natural areas, national parks, and wilderness areas, has increased substantially since the MLU series began in 1945. Urban land use has also modestly increased, as population and economic growth spur demand for new housing and other forms of development.[6]

Land use change happens for many reasons, including changing timber prices, new or revised natural resource policies, urban pressure, and more.[7]

Land use planning regulates usage and is intended to avoid land use conflicts, although conflicts still occur. Federal, state, and local governments regulate growth and development through laws; two major federal laws significantly limit land use: the National Historic Preservation Act of 1966, intended to preserve historic and archaeological sites in the United States; and the National Environmental Policy Act of 1969, which established the President's Council on Environmental Quality.

Still, the continuing transformation of land around the world ranks as one of the biggest potential threats to human health and environmental sustainability.

Negative Impacts

- **Habitat fragmentation.** This occurs when a large tract of land habitat is split into smaller patches. This patchwork can have a severe impact on wildlife. As habitat is reduced, it can lead to increased competition among species for limited resources.
- **Deforestation.** Deforestation is the mass destruction of trees. Forests cover 30 percent of the planet, but are disappearing at an alarming rate due to forest clearing for agriculture (crops or livestock), logging, and infrastructure expansion such as road building and urbanization. Since 1990, "the world has lost about a billion acres of forest," according to the Food and Agriculture Organization of the United Nations.[8] "About 17 percent of the Amazonian rainforest has been destroyed during the past fifty years. Eighty percent of Earth's land animals and plants live in forests, and deforestation threatens species including the orangutan, Sumatran tiger, and many species of birds. Removing trees deprives the forest of portions of its canopy, which blocks the sun's rays during the day and retains heat at night," leading to extreme temperature swings.[9]

- **Heat islands.** These are pockets of warm air that occur when trees and other vegetation are replaced by human-made materials such as concrete or buildings.
- **Agriculture.** Agricultural land uses such as plowing, the method of plowing, livestock grazing, manure spreading, and pesticide and fertilizer applications can affect water quality and watersheds. Agricultural land use may replace native habitats, and open areas can lead to erosion and dust.
- **Invasive species.** Some land use invites invasive species. Invasive species are not native to a particular location and cause damage to ecological communities when they move in. They can displace and even cause the extinction of native plants and animals. They reduce biodiversity by outcompeting natives for limited resources, including habitat.
- **Runoff.** Nonpoint source water pollution is caused by rain or melting snow moving over usually impervious landscapes. As the runoff moves, it picks up pollutants, eventually depositing them into waterways and groundwater.[10]

Nonpoint source pollution can include:

- Fertilizers, herbicides, and insecticides.
- Oil and toxic chemicals.
- Sediment from construction sites and eroding streambanks.
- Salt from treating roads in the winter.
- Waste from animals and septic systems.
- Atmospheric pollution that falls to the Earth in the form of snow and rain.
- Point source discharges include water used in industrial and municipal facilities.

"States report that nonpoint source pollution is the leading remaining cause of water quality problems. The effects of nonpoint source pollutants on specific waters vary and may not always be fully assessed."[11]

Obstacles

- **Lack of data.** The National Research Council has identified land use dynamics on a global scale as an important area for developing environmental research. There is too little data about national trends in land use and potential human health impacts.
- **Data classification and measurements.** These vary—comparing apples to oranges. Some data collection is particular to an industry. Some data relies on a particular collection method or categorization that makes it a challenge to compare and integrate the data over time.
- **Variety of regulations.** There are myriad federal, state, and local regulations that govern land use.

Applying Critical and Creative Thinking

When we apply critical and creative thinking to land use discussions, it is helpful to ask ourselves some questions: Do I know the history of the land that I live on? Have I explored my neighborhood and talked to neighbors who have lived there longer than I have? What changes has my community experienced in terms of land use? Do I have access to a park or public green space or know where to find that information? What critters and plants share my green space?

Communities of all sizes are getting creative when it comes to preserving, protecting, and redeveloping land.

Smart Growth

"Smart growth" includes strategies aimed at protecting the environment, while making communities more diverse, commercially viable, and attractive. The Smart Growth Network is a partnership of government, business, and civic organizations that supports smart growth, and offers these suggestions for smart growth strategies when planning land use:

- "Mix land uses.
- Take advantage of compact building design.
- Create a range of housing opportunities and choices.
- Create walkable neighborhoods.
- Preserve open space. . .
- Make development decisions predictable, fair, and cost-effective.
- Encourage community and stakeholder collaboration in development decisions."
- Reinvest in existing infrastructure and rehabilitate historic buildings.
- Design neighborhoods that have homes near stores, offices, schools, houses of worship, parks, and other recreational sites. Make it easy to walk, bike, and take public transportation. This also creates business opportunities and supports the local tax base.
- Support housing that makes it possible for senior citizens to stay in their neighborhoods as they grow older. Create funding options and offer resources to support those interested in buying their first home. When there are people with similar needs—such as those with disabilities and seniors—consider a housing coop to share services.[12]

Recognize That Not All Communities Face the Same Land Use Crisis

During the past several decades, tribes, corporations, and the government have fought over control of Native land and resources. Earlier, we discussed treaty rights as they relate to fishing. Here, we look at highly visible protests and legal proceedings related to pipelines

Scores of tribes traveled to North Dakota in 2016 to protect ancestral homelands compromised by the Dakota Access pipeline. The pipeline runs from western North Dakota to southern Illinois. It snakes beneath the Missouri and Mississippi Rivers and under part of Lake Oahe near the Standing Rock Indian Reservation. Many tribal and nontribal community members consider the pipeline to be a serious threat to the region's water, their history, and their culture.

The protesters came to speak for their ancestors who are buried on those sites. They came to speak for their right to preserve their ceremonial sites and drinking water. And they came to speak for the subsequent generations who will have to live with the results of the pipeline's construction.

In March 2020, a US District Court judge ruled that the government had not sufficiently "studied the pipeline's effects on the quality of the human environment." The judge ordered the US Army Corps of Engineers to conduct a new environmental impact study.

In July 2020, a US District Court judge ruled that the pipeline should be shut down while a new environmental review was conducted. But one month later, the temporary shutdown order was overturned by a US appeals court.

In 2021, protesters tried to block pipeline and energy company Enbridge's construction efforts that would expand and repair a controversial pipeline that would carry hundreds of thousands of barrels of oil through tribal lands and watersheds in northern Minnesota.

Then on his first day in office, President Joe Biden denied the company a key permit. So after more than ten years of protest and drawn-out legal battles, the Keystone XL pipeline, backed by the Canadian energy infrastructure company TC Energy, abandoned the project in June 2021.

While these protests may have resulted in different outcomes, in many ways, the protesters in each of these pipeline cases have been fighting for the land, which cannot fight for itself.

As wrote Paula Gunn Allen, Laguna Pueblo,

> We are the land . . . that is the fundamental idea embedded in
> Native American life[,] the Earth is the mind of the people as

we are the mind of the earth. The land is not really the place (separate from us) where we act out the drama of our isolate destinies. It is not a means of survival, a setting for our affairs . . . It is rather a part of our being, dynamic, significant, real. It is our self.[13]

Selecting and Applying Action Steps

- Spend time gardening. Every year our family garden presents surprises. We evaluate the successes and the failures—but sometimes the reasons for the results remain mysteries.
- Go camping with a young person who hasn't had the opportunity, a mentor, or access to outdoor recreation.
- Dig your own worms for fishing.
- Go canoeing or kayaking and see the land from a different vantage point.
- Try bird-watching and get to know the critters with which you share the land.
- Have a picnic. Play in the park.
- Go for a hike—up a mountain, on the beach, or at a nearby nature reserve.
- If you are not an outdoors person, you can start small by spending time in the yard or planting just one tree.
- Think of land as a playground. What activities can you do to enjoy the space you have?

Benefits of Walking

"In every walk with nature one receives far more than he seeks," wrote John Muir, one of America's most famous naturalists and conservationists.

"Walking is my favorite form of exercise. I do a lot of thinking when I walk. I talk to neighbors and feel a sense of community. I notice the seasonal changes occurring in my neighborhood. I stop to pick up trash for proper disposal when I return home. Sometimes I am joined by my dog, who loves to take in the scents and stops to listen for other pups. If I pass a Little Library, I might pause to take a peek inside and even sample some of the prose. If I find a bench in the park, I might sit for a bit and simply stare at the pond across the path and marvel at its inhabitants.

"A trail is not for hurrying. Despite how familiar the trail is to me now, there is always something new to see or hear as I wind my way along."[14]

Walking allows:
- A chance to experience the changing seasons.
- Time to notice the nuances of our surroundings.
- An opportunity to interact with our neighbors—the human and animal ones.
- To feel the ground beneath us and to keep moving forward.
- To be more fit.

It's not always about how fast you get there—but that you enjoy your journey.

Land Is Dynamic

"Land is much more than an inert bit of soil upon which we can erect a building, create a parking lot, or even grow some leaf lettuce in the backyard garden. It is alive, filled with organisms of every shape and kind."[15]

Like all of us, land has a history, and land use by humans has a long history.

It is important to understand the long-standing history that has brought you to reside on the land, and to seek to understand your place within that history. Land acknowledgments do not exist in a past tense, or historical context: colonialism is a current ongoing process, and we need to build our mindfulness of our present participation. It is also worth noting that acknowledging the land is Indigenous protocol.[16]

Know Your Roots

Where we grow up and where we live today is a measure of who we are and how we got here.

When Jerry was a child, the home farm included 160 acres. The fields had been farmed since the 1850s and the area—Wild Rose, Wisconsin—was where the last great glacier stopped distributing rocks of various sizes and colors.

About 100,000 years ago, the Laurentide Ice Sheet spread across the continent. As the cycle was coming to an end, starting about 31,500 years ago, the glacier began to move into Wisconsin. It took another 13,500 years before temperatures warmed and the glacier began to melt and retreat from Wisconsin.[17]

Glaciers created much of the Wisconsin topography with the exception of what is now dubbed the Driftless Area in southwestern Wisconsin. "Ice movement churned the ground and created rich soil, which is key to our farming heritage. In fact, without the last glaciation, Wisconsin might not be the dairy state!"[18] And Jerry's family might not have been farmers.

CHAPTER 12

Endangered Species

Going, Going, Gone

Background

On our farm near Wild Rose, Wisconsin, we take pride in one of our tiniest—yet most colorful—residents, the Karner blue butterfly.

The thumbnail-size butterfly has a lifespan of just a few days to a few weeks and was listed federally as an endangered species in 1992. Its numbers saw a sharp decline after extensive habitat destruction and the loss of its only food supply: the wild lupine. Karner blue caterpillars only feed on wild blue lupine leaves, while the adults feed on the flower's nectar. This diet severely restricts where Karner blues can survive.

Wild blue lupines' habitat includes pine barrens, oak savannas, and lakeshore dune habitats. They prefer sandy soils and sunny open patches. These habitat types also often support other rare species.[1] The wild lupine is part of the legume family—yes, it's related to peas—and we scattered the pods from the wild lupine to help seed the sandy soil of our farm. Lupine also helps to increase soil nitrogen.

"By the time the Karner blue was listed as endangered in 1992 the number of Karner blue butterflies had declined by at least 99%."[2]

Unfortunately, the story of the Karner blue is not unusual. The US Fish and Wildlife Service maintains a list of species, such as the Karner blue, that are federally endangered and threatened. The list in the United States includes 1,675 species of insects, reptiles, mammals, fish, and more.[3]

The Environmental Problem

Endangered species are those animals and plants that are in danger of becoming extinct—wiped out, erased, gone forever, just like the woolly mammoth, dodo, great auk, Tasmanian tiger, and passenger pigeon.

The last passenger pigeon, Martha, died in the Cincinnati Zoo in 1914. Once endemic to North America, passenger pigeons flocked together in such huge clouds that historic records recall them blocking out the daytime sky for hours at a time. In the late nineteenth century, the passenger pigeon was hunted for food and because they were believed to be a threat to agriculture. Their populations also thinned as large forest areas were cleared for development across North America.

Threatened species are animals and plants that are not yet at the finish line, but on the path of the passenger pigeon, and are likely to become endangered in the foreseeable future.

Neither is a good way to be categorized. But worse? Extinction.

In September 2021, the US Fish and Wildlife Service declared the ivory-billed woodpecker and twenty-two more birds, fish, and other species extinct after exhaustive efforts to try to find any living members of the species failed.

Worldwide, 902 species have been documented as extinct during the last five centuries, joining the likes of the Baiji white dolphin and West African black rhino. However, "the actual number is thought to be much higher because some are never formally identified, and many scientists warn the earth is in a crisis with flora and fauna now disappearing at 1,000 times the historical rate."[4]

The loss of both animal and plant species is occurring at such an alarming rate that some scientists argue we are on the verge of the sixth mass extinction. While the other five mass extinctions were the result of natural geological causes, such as volcanic activity, the current mass extinction, scientists argue, is being caused by humans.[5]

Paleontologists characterize mass extinctions as "times when the Earth loses more than three-quarters of its species in a geologically short interval, as has happened only five times in the past 540 million years or so. Of the four billion species estimated to have evolved on the Earth over the last 3.5 billion years, some 99 percent are gone."[6]

The Five Mass Extinctions

1. **End Ordovician** (444 million years ago [mya])
2. **Late Devonian** (360 mya)
3. **End Permian** (250 mya)
4. **End Triassic** (200 mya)
5. **End Cretaceous** (65 mya)[7]

Today, the reasons for extinction run the gamut. Government scientists cite climate change at the top of the list. But urban development leads to habitat fragmentation and loss. Air and water pollution impact plant and animal species at a variety of levels. There is increasing competition from invasive species. Poachers are targeting certain species, such as the Asian elephant, for economic gain. Hunting methods and tools have become more sophisticated, making it easier to overharvest.

Damming of rivers, water pollution, and invasive species are threatening aquatic ecosystems worldwide. Coral reefs are declining.

Like the Karner blue butterfly, monarchs are particular about what they eat and where they live. Milkweed is essential for monarchs and is the only plant on which they'll lay their eggs and the only one that their caterpillars will eat. But in the last two decades, development and herbicide use have depleted monarch habitat.[8]

In 2022, the monarch butterfly, known for its migratory prowess, was officially designated as endangered by the International Union for Conservation of Nature, with weather impacts, loss of milkweed and habitat, and pesticide use cited as contributors to population decline.[9] We have monarchs on our farm in Wild Rose, just as we have Karners. That's because we have milkweed and lupines. This link between plant species and pollinators is critical to both the butterfly's success and one that we heavily consider when planning our habitat management.

The loss of species crosses every taxon, but the highest rate is occurring among amphibians due to "habitat loss, water and air pollution, climate change, disease and more. Scientists estimate that a third or more of all the roughly 6,300 known species of amphibians are at risk of extinction."[10]

Obstacles

- Who cares? Some wonder, "with 7.8 million animal species on the planet, what difference does losing a few thousand make?"[11]
- Misleading or false sources of information. People have a wealth of information available to them. Not all of that information is accurate. In fact, there are sources deliberately misleading the public. These sources include individuals with a financial gain to be made from misleading others to special interest groups or organizations that have a particular agenda.
- Lack of data. Sometimes we don't know what we have before it is too late. In 2000, fewer than 20,000 species had been evaluated for population.[12]
- Zoos are saving and breeding endangered animals, so let's not worry about the ones in the wild. Association of Zoos and Aquariums–accredited spaces are working to save some endangered animals through captive breeding, species reintroduction

to wild spaces, habitat conservation, education, and more. But many zoos aren't focused on replenishing threatened populations. Rather, they are intent on getting visitors through the gates. Seeing an endangered animal in captivity might give someone the false sense that that species is going to recover. Yet not all species respond well to a life in captivity.

- Inability to develop land if it is designated as critical habitat based on the presence of an endangered species. When a property is designated as critical habitat, federal agencies need to ensure that the actions the landowner plans to take during development "do not destroy or adversely modify that habitat."[13]

Applying Critical and Creative Thinking

Rachel Carson's 1962 book, *Silent Spring*, shared the sad truth that bird populations across the United States were declining as a result of widespread application of the pesticide DDT (dichlorodiphenyltrichloroethane). Because of its impactful message, *Silent Spring* has been credited with launching the modern environmental movement.

Carson shared research showing how birds ingesting DDT tended to lay thin-shelled eggs that would then break in the nest. DDT was in fact largely to blame for bringing bald eagles, peregrine falcons, and some other bird populations to the edge of extinction.

Since *Silent Spring* was published, bald eagle numbers have rebounded, and the fact that bald eagles soar throughout the United States today is the result of years of protection under the Endangered Species Act and conservation measures such as banning the pesticide DDT in 1972.[14]

In *Silent Spring*, Carson wrote, "One way to open your eyes is to ask yourself, 'What if I had never seen this before? What if I knew I would never see it again?'"

Case Study
The Northern Spotted Owl/Old-Growth Forests

The high-profile and decades-long controversy over the northern spotted owl in the Pacific Northwest has drawn attention to the question of whether society should protect animals from extinction. Today, the spotted owl is a symbol for saving old-growth forests and as such, is driving discussions over whether saving a particular habitat should take precedence over a community's economic growth.

The northern spotted owl is, like our little Karner blue butterfly, listed as federally endangered and has received protection under the Northwest Forest Plan. The plan, which was put in place in 1994, seeks to restore habitat that is necessary for the northern spotted owl, as well as the marbled murrelet (a bird that requires old, mature forest habitat for nesting), and salmon stocks.[15] Need another reason to love owls? Owls are predators—and darn good ones—and help control rodent populations.

Sounds like a species that deserves saving, right? But saving the owls' habitat could cost jobs. And the timber industry has argued that cutting the old growth is essential for providing people with the wood and paper products they need. They further argue that millions of acres of old-growth forest have already been set aside as national parks and wilderness areas. They add that logging sites can be replanted.

"Even as unemployment in some timber counties routinely rises into double digits, there are no longer presidential Timber Conferences, like the one Bill Clinton held in Oregon in 1993 seeking middle ground between conservation and protecting rural economies. Many factors contribute to rural declines, but logging restrictions played a role."[16]

And the logging industry isn't the only threat to the spotted owl. The other main threat is competition with barred owls, a species that is widespread throughout the eastern United States but also overlaps with the range of spotted owls in the Pacific Northwest.[17]

How do we reconcile these seemingly divergent views on support for and against spotted owl protection? Is this just a species caught up in what has become the politics of protection?[18]

Not having a clear and immediate answer to these questions should not preclude thinking critically and creatively about them. What about providing timber companies incentives to create spotted owl habitat? These could include tax incentives, forest conservation easements, and carbon tradable credits based on the amount of carbon sequestered by the forest habitat.

Can our forestry practices be as diverse as the species inhabiting the forests—perhaps use thinning of trees over clear-cutting? Can we build ecotourism in areas where jobs are needed?[19]

Maybe a starting point is to consider that the spotted owl is an indicator or keystone species, meaning that its health reflects the health of the ecosystem it is a part of. In this case, old-growth forest.

Perhaps those who have not spent time in an old-growth forest should give it a try.

"Thus public apathy toward older forests may constitute the most serious threat to their continuation. Data are scarce on the feelings most people have about older forests, and contact between the average citizen and older forests is decreasing with increasing urbanization. It will be important to understand people's attitudes toward older forests and to make a factual and compelling case for retaining and restoring them."[20]

Achievements such as the Endangered Species Act remind us of our responsibilities as humans and how such actions can promote the continuation of life on our planet. There are many lessons we can learn from the kind of critical and creative thinking the act created, and many examples of the fruits they bear that we can observe.

- You may argue that there is a moral argument for protecting other species. The Endangered Species Act doesn't claim that human existence depends on the existence of wild species. But it does say that we don't let species go extinct because that caring and feeling—the ability to think and use logic—is part of what makes us human. When another species needs us, we will go to their aid.

- Natural products from plants and animals save lives. According to one 2012 paper published in the *Journal of Advanced Pharmaceutical Technology & Research*, "Nearly half of all drugs created within the last 30 years are either directly or indirectly from natural products."[21]
- Bees and butterflies are essential to pollinate your plants, including food sources.
- Ecotourism generates billions every year throughout the world—bird-watchers alone spend nearly $41 billion on travel and equipment.[22]
- Plants and animals can be the "canaries in the coal mine." Freshwater mussels, for example, filter water, and their disappearance can signal water pollution issues. The declining eagle and peregrine falcon populations drew attention to the dangers of DDT.

Selecting and Applying Action Steps

- Save and restore habitat. We can protect endangered species by protecting the wild places that they call home.
- Encourage and engage in coordination among federal, state, tribal, and local officials. Animals and plants cross borders, and so should collaborations.
- Be patient, but diligent. Helping a species rebound—maybe move from the threatened list—takes time and commitment that is often dependent on habitat and food availability, reproduction rate, and climate.
- Promote and be a reliable source of information.
- Look for connections. The loss of a species can result in spiritual and cultural losses as well. In Asia, for example, people and elephants have had a special relationship, and elephants are revered as cultural icons.

- Minimize use of herbicides and pesticides. Reduce use of fertilizer.
- Choose biological pest controls over chemical ones. Some critters can be safe and effective ways to manage a pest problem.
- Americans value open spaces and diversity in wildlife and habitat. Many try to live in ways that do less environmental harm, and they seek protection of wild spaces and species that allows us to pass this heritage along to our children and grandchildren.
- Find and share positive examples of survival due to human interventions—as was the case with banning DDT to bring back healthy populations of peregrine falcons and bald eagles.

The Government Stepped In: A Rare Bipartisan Agreement

In response to the crisis facing plant and animal species across the planet, the Endangered Species Act was passed with bipartisan support in 1973. This landmark law allows consideration of listing a species as endangered or threatened, thereby affording it protections.

Each petition undergoes scientific evaluation and public review. The law requires recovery plans and habitat protection for listed species.

Endangered populations are then monitored to determine whether a given species is recovering. A species is removed from the list when they are considered recovered.

The act also supports species conservation outside of the United States "and is the law through which the United States enforces the Convention on International Trade in Endangered Species (CITES). CITES is a global agreement to monitor, regulate, or ban international trade in species under threat and is a key tool in the fight against the illegal wildlife trade."[23]

Congress amended the act in 1978, 1982, and 1988.

One story of success using science to inform critical and creative thinking, and then moving to action, comes from the Isle Royale National Park in Michigan. The island's most famous residents are its wolves and moose. Top predators, such as the gray wolf, are very important to healthy ecosystems.

The relationship between the island's wolves and moose has been the subject of a study that began in 1958, making it the longest predator-prey study in the world.

In 1980, researchers saw a decline in wolves on the island after a virus introduced by a hiker's dog spread throughout the wolf population. Another suspected reason for a drop in wolf numbers was that historically, ice bridges had allowed wolves to travel over Lake Superior. But as the ice bridges formed less frequently, the wolves were no longer able to migrate.

By 1980, there were just two wolves left on the island. Without wolves, the moose population surged, and that spelled danger for the island's vegetation. The moose population increased to more than 1,500, and they were consuming a lot of the native plant species, including aspen, birch, balsam fir, and Canada yew. To reestablish balance, the National Park Service decided to introduce twenty to thirty new wolves to Isle Royale. The plan was based on scientist and public input.

The Park Service and the State University of New York–College of Environmental Science and Forestry released a study in December 2019 showing that the plan was working and that the new wolves were hunting moose.[24]

Some had argued for a moose hunt to control the herd instead of reintroducing wolves. The Park Service, however, rejected moose hunting in 2018, noting that results of the hunts would be unpredictable, would require extra staffing to monitor the harvest, and that the whole process was logistically burdensome.

The story of Isle Royale wolves shows that if you remove one animal or plant species, it may upset the balance of species and even change the ecosystem completely.

Back to That Little Blue Butterfly

Bees and butterflies may seem, to some, insignificant or even pesky. They are small and not as physically charismatic as the moose on Isle Royale or the American bison.

Bison, North America's largest land animal, were nearly driven to extinction by habitat loss and hunting. An estimated 30 to 60 million bison roamed North America until the late 1800s, when their numbers dwindled to fewer than 1,000. Today, after efforts were put in place to protect the bison and their habitat, about 30,000 bison exist in herds managed by government and conservation organizations.

Small but mighty, butterflies like the Karner blue play a huge role in our ecosystem—they are pollinators. To manage for Karner blue butterflies, it is essential to also manage for their habitat. Oak savanna and pine barrens ecosystems that Karners depend on are threatened. Recovering the Karner means recovering these ecosystems that support it.

Butterflies are not only beautiful, they are symbolic and often used as a metaphor for representing transformation, change, hope, and life. These are all important elements to think about if you are, for example, working to save—or share an address—with an endangered species such as the Karner blue.

CHAPTER 13

Biodiversity

Protecting Healthy Ecosystems

Background

"This would never happen on a normal lake, because a normal lake is knowable. A Great Lake can hold all the mysteries of an ocean, and then some," Dan Egan writes in the introduction to his *New York Times* best-selling book, *The Death and Life of the Great Lakes*.[1]

Egan, a two-time Pulitzer Prize finalist, *Milwaukee Journal Sentinel* journalist, and senior water policy fellow at the University of Wisconsin–Milwaukee's School of Freshwater Science, is referring in the introduction to the ecological catastrophe happening to the Great Lakes.

The Great Lakes form the largest surface freshwater system in the world. Together, lakes Michigan, Superior, Huron, Erie, and Ontario contain nearly one-fifth of the Earth's surface fresh water, the region that sustains a rich array of ecosystems and more than 6,000 plant and animal species, some of which are found nowhere else on Earth. "The Great Lakes have over 10,000 miles of shoreline and serve as a drain for more than 200,000 square miles of land, ranging from forested areas to agricultural lands, cities, and suburbs," according to the National Wildlife Federation.[2]

The Great Lakes watersheds are home to some of North America's most majestic wildlife, including the gray wolf and bald eagles. The lakes are home to an abundance of fish species including whitefish and trout. Migratory birds use the area as a stopover to rest and fuel up during their spring and fall migrations. The Great Lakes region houses a variety of waterfowl, and provides important breeding and rest-over areas for the common loon and endangered Kirtland's warbler.

The region is massive and biologically diverse, and, as Egan writes, is in trouble. The region is at risk from climate change, pollution, and invasive species.

The diversity of life, like that found—and not endangered—in the Great Lakes region, is impressive, and this biodiversity is also something to be celebrated. And it is. The United Nations celebrates May 22 as the International Day for Biological Diversity, with the goal of educating on the topic to increase understanding and awareness of biodiversity issues.

"Biodiversity," combining the word "biological" and "diversity," is defined as "the variety of life in the world or in a particular habitat or ecosystem. It can be used more specifically to refer to all of the species in one region or ecosystem."[3]

"Each higher organism is richer in information than a Caravaggio painting, a Bach fugue, or any other great work," wrote Professor Edward O. Wilson, often called the "father of biodiversity," in his seminal 1985 paper "The Biological Diversity Crisis."[4] Wilson went on to add, "If astronomers were to discover a new planet beyond Pluto, the news would make front pages around the world. Not so for the discovery that the living world is richer than earlier suspected, a fact of much greater import to humanity."[5]

You might not have heard of Wilson before, but he discovered that ants communicate by transmitting chemical substances known as pheromones. He wrote twenty books, won two Pulitzer Prizes, and discovered hundreds of new species.

Today, more than "1.7 million species of animals, plants and fungi have been recorded, but there are likely to be 8 to 9 million and possibly up to 100 million."[6] The motherlode of biodiversity is the tropics, a warm and moist area teeming with species. Still, despite 250 years of taxonomic classification, research suggests "that some 86 percent of existing species on Earth and 91 percent of species in the ocean still await description."[7]

If you include bacteria and viruses, the number might soar into the billions. "A single spoonful of soil may contain 10,000 to 50,000 different types of bacteria."[8]

Biodiversity may be grouped into levels, starting at the gene level, then individual species, communities, and ecosystems. Areas such as Brazil and Madagascar, with high levels of biodiversity, are called hotspots. Biodiversity is usually broken into four areas of diversity: genetic, species, community, and ecosystem.

Types of Diversity

Genetic Diversity

Genetic diversity refers to how closely related the members of one species are in a particular ecosystem. If all members have similar genes, there is low genetic diversity. Low genetic diversity may occur when there is inbreeding, which can also lead to passing along undesirable traits or increasing susceptibility to disease. Having a higher level of genetic diversity is preferable because then a species is more likely to be able to adapt to changes in its environment.

What happens if genetic diversity is severely reduced? The Food and Agriculture Organization of the United Nations has warned, for example, "that declining genetic diversity in food and agriculture makes food crops and livestock more susceptible to disease and farmers more vulnerable to crop failure."[9]

Species Diversity

When an ecosystem has a large number of species, it is considered to have species diversity.

Community Diversity

A natural community is one where plant and animal species share a habitat. Communities usually are named after the dominant plant species found there (examples include pine barrens and oak savannas). Communities can be as small as an acre to as expansive as thousands of acres.

Ecosystem Diversity

The degree of ecosystem diversity in an area is based on the variety of habitats, communities, and other resources that are present. A region composed of several ecosystems—high ecosystem diversity—may be better equipped to support species survival when one ecosystem is threatened by drought or disease.[10]

Forests are one of the most biodiverse terrestrial refuges. Forest biological diversity includes trees, of course, but also the understory and plants, animals, and microorganisms that inhabit the forest area and live in the canopy. Forests contain "80 percent of amphibian species, 75 percent of bird species and 68 percent of mammal species," according to *The State of the World's Forests*.[11]

The Environmental Problem

"A 20 percent drop is widely considered the threshold at which biodiversity's contribution to ecosystem services is compromised. It's estimated that over a quarter of Earth's land surface has already exceeded this threshold."[12]

The biggest threats to biodiversity include:

- **Pollution.** Pollution—in some cases, even in small amounts—can negatively impact plants and animals. It can lead to extinction.

- **Habitat Destruction and Fragmentation.** Habitat loss to development and filling in wetlands are serious contributors to species population declines. Certain human activities lead to fragmentation that creates habitat spaces that are too small to support some animals. Aquatic ecosystems also can become fragmented when they are filled and modified.

- **Invasive Species.** The introduction of a nonnative species to an ecosystem can crowd out native species, create competition for resources, and even cause extinctions. Not all nonnative species cause a problem when introduced to a new environment, but if that species causes harm, then it's called a nuisance or invasive species. Invasive species are often introduced through human activity such as transporting materials. Some are intentionally introduced, such as when someone decides they no longer want an exotic pet and releases it into a local habitat.

- **Illegal Collection and Hunting.** Some animals are poached, such as elephants for the ivory trade, and some are illegally collected for the pet trade.

- **Changes in Climate.** Some species are unable to adapt to changes in the climate, including increasing temperatures and rising sea levels.

Obstacles

- There are significant gaps in our knowledge of diversity. (Although not easy, there are ways we can measure it, ranging from measuring genetic diversity to documenting extinction rates.)[13]
- There is a perception that loss of biodiversity is unavoidable.

- Some consider any disturbance, especially fire and clear-cutting, as always being detrimental. However, prairies, for example, rely on fire for species diversity to occur.
- Some believe that the natural world is better left alone and will heal itself. But sometimes we have gone too far for that to happen, and human intervention is often necessary. For example, what would happen to a prairie today if we decided to just leave it alone? Would it be able to rebound in the absence of prescribed burns? That is unlikely.
- Politics. Legislation is needed to support wetland protection and restoration, for example. While there is growing appreciation for the roles wetlands play in flood mitigation, sediment and nutrient filtering, and groundwater recharge, the laws have not kept pace with that appreciation.

Applying Critical and Creative Thinking

Environmental journalist and author Elizabeth Kolbert is perhaps best known for her 2014 Pulitzer Prize–winning book, *The Sixth Extinction: An Unnatural History* (2014), which details the crushing of biodiversity beneath the human footprint. She writes in the prologue, "Very, very occasionally in the distant past, the planet has undergone change so wrenching that the diversity of life has plummeted."[14] The book's title refers to the big five prehistoric mass extinction events as well as the current feared sixth and first human-caused mass extinction. In her book, she paints a picture of a damaged world and suggests humans are changing the world, perhaps without even realizing they are doing it. She calls humans a "Faustian" species. Faustian is defined by Merriam-Webster as "made or done for present gain without regard for future cost or consequences."[15]

If biodiversity refers to every living thing, how can we engage people in an environmental problem so big? Maybe reciting a list of factoids isn't the answer. We want facts—the truth to inform our thinking. But we need something more to keep the scientifically credible conversation about something as big as biodiversity from collapsing under its own weight.

In his 1996 book, *The Idea of Biodiversity*, David Takacs interviewed prominent conservation biologists and "revealed a telling dichotomy between the reasons that inspire them to care about biodiversity, and the reasons they provide when trying to convince others to care. Inspiration often comes from an emotionally driven love of nature."[16]

So, maybe we also need to get creative when broaching the topic of biodiversity. Perhaps we speak to our stomachs and the impacts on the food chain if, for example, pollinators are eliminated. Maybe we start small—with the tiny beings—the insects such as the bee.

Perhaps, for example, we grab our audience's attention with the engineering lessons learned from some insects—spiders and ants. Yes, small critters that make some cringe. But it's a fascinating fact that self-driving car algorithms have been inspired by Amazonian ants.[17] Some researchers are studying spider webs to learn how to build stronger cables. What about weeds or molds—why are they important? You can defend them as useful to human life. Molds yielded penicillin.

Then there is the lure of mystery. Who doesn't love a great mystery novel? Or the draw of space and what lies beyond our galaxy?

One concern about losing biodiversity is that many species are being lost before we even know they exist or how important they are to the rest of the ecosystem. Take the ocean, one of the main repositories of the world's biodiversity. It contains some 226,000 known species and many more remaining to be discovered.[18] We literally have not reached the depths of discovery that are possible in the ocean or in soils.

We are just beginning to understand the diversity of the microbiome, where there are both good and bad actors. "The microbiome is defined as a community of microorganisms (such as bacteria, fungi, and viruses) that inhabit a particular environment and especially the collection of microorganisms living in or on the human body. 'Your body is home to about 100 trillion bacteria and other microbes, collectively known as your microbiome.'"[19]

Selecting and Applying Action Steps

- **Learn about biotic inventory methods and practice those.** One way to do that is to participate in a BioBlitz, an event during which the goal is to find as many species as possible in a specific area during a specific place in time to provide a snapshot of an area's biodiversity. These events can happen anywhere—even in areas as small as a backyard or a neighborhood park.[20]

- **Support sound scientific evidence.** Anyone can do this by contributing to the Global Biodiversity Information Facility, a free and open-access biodiversity database. Data here includes everything from museum specimens collected hundreds of years ago to geotagged smartphone photos.[21]

- **Learn from the past.** Knowledge of presettlement vegetation teaches us the ecological potential of an area and suggests where biodiversity might be restored. An example would be the big bluestem that we have on our farm. It is the characteristic plant species of the North American tallgrass prairie. From past data, we can surmise what conditions—from soil to climate—that it needs to thrive. We can then work to replicate those presettlement conditions to support big bluestem growth today.

- **Use appropriate genotypes in restoration and management.** This includes native seed banks where seeds are stored to preserve genetic diversity and for future reseeding. Globally, it's estimated that 40 percent of plant species are vulnerable to extinction. Seed bank vaults are usually flood, bomb, and radiation proof, and have temperature and humidity controls. More than 1,000 seed banks exist worldwide. Some have a particular focus.[22]

- **Tout health benefits.** Biodiversity indicates the health of an ecosystem. Biodiversity provides resources for medicines and the potential for future cures for diseases.

- **See connections.** Losing a single link in the natural network can have a domino effect. The loss of one species can disrupt an entire food chain. Bees, for example, are essential plant pollinators, but due to pesticide use, their numbers have declined. The No Mow May initiative was made popular by the UK organization Plantlife and is gaining momentum in the United States. The goal of No Mow May is to allow grass to grow unmown during May, creating habitat and forage for early season pollinators.[23]

- **Promote sustainability.** Sustainability is a biological system's ability to remain diverse and thrive over time. A long-lived forest is an example. Sustainability requires us to balance environmental, social, and economic demands, which are also referred to as the "three pillars" of sustainability.

- **Money as a measure.** Efforts to preserve biodiversity will be more successful if they address the economic and cultural factors that drive deforestation and development. "If money is a measure, the services provided by ecosystems are estimated to be worth trillions of dollars."[24]

- **Spark ingenuity.** One promising initiative is an open-source genetic database for all plants, animals, and single-cell organisms on the planet.[25]

Conclusion

Just the Beginning

While we have reached the end of this book, it really is just the beginning. From here, it is up to each of us to consider and adopt those strategies most applicable and reasonable to us, to—if you accept our challenge— help us create an environment where good ideas can grow.

We hope we have interested you in thinking about tough and sometimes ethical questions that together we can make progress by finding answers to, and that in doing so, we can find practical and affordable solutions to some of the most serious environmental problems.

In this book, we have outlined the history of the environmental movement shared the components of critical and creative thinking, and explored contemporary environmental debates, showing how they translate to, or become entangled in public policy. We have illustrated the importance of bipartisanship with examples of successful environmental programs such as Wisconsin's Knowles-Nelson Stewardship Fund and the federal Endangered Species Act. We also have explored historical legacies such as the move from a rural to a more urban lifestyle, changes in technology, challenges to treaty rights, significant changes in land use, effects of pollution, depletion of natural resources, and dwindling biodiversity. Our ecological crisis, climate change, forces us to consider the strained relationship between science and politics, and what to do to move ahead together to create positive change. It forces us to see the urgency for action.

We suggest that what is clear and can be a common starting point for all of us is the fact that what we are doing right now is not working for the planet or for people as a whole. Whether you believe, as some argue, that we are facing the sixth mass extinction, what is undeniable is that human behavior is impacting the global environment in some very negative and disastrous ways. People are paying a high price for this behavior. We know that climate change is causing tremendous harm, and we are not equal in the face of the climate challenge. In fact, it is the poorest areas that often are most affected by uncontrolled global warming.

If we can agree, then, that climate change is a moral issue, aren't we obligated to action to right the wrongs? Most agree that we need to act today to create a more healthy, sustainable, and equitable world for future generations. We have cited the Seventh Generation philosophy, an ancient Iroquois philosophy that is held by many Indigenous people around the world, which tells us we must consider the impact of our decisions on the next seven generations.

But in our research, we find little clarity or agreement based on our obligations or on what exactly they are. We find that nations, states, and municipalities lack consensus on what to do next. Many lack a forum for discussion or enough respect for one another to listen to diverse viewpoints. An inability to, or interest in, sifting fact from fiction is a crisis.

In this book, we offer an interdisciplinary approach to thinking about the environment with nods to social economics and physical and mental health, along with examples from areas such as agriculture, air quality, biodiversity, forestry, and water resources. In each chapter, we seek to show that the health of the planet and people are connected.

We point to links between environmental, social, and racial justice and emphasize that healthier environments for underprivileged communities requires access to clean air, water, and soil. Who causes pollution and who is most likely to suffer from that pollution? That climate change disproportionately falls on those who are least able to prepare for, and recover from natural disasters.

We have called the book *Planting an Idea*. As with most plants, most ideas need space and time to grow. But we are concerned that humanity has lost its patience. We expect immediate results and satisfaction. We jump on drama posted to social media and reality TV, and are fueled by fake news. Maybe, we would rather be numbed than nudged. With this book, we want people to stop and think. That means listening and creating venues for open and honest conversation. That means meeting people where they are and learning how they got there.

Communication is key. Too often, we fear criticism and we keep our ideas to ourselves. An idea may need time to evolve and grow. But to do so, it needs feedback. What we need is to make meaningful connections even with those with whom we disagree. We need to ask questions, observe, and seek the truth. We can, and must, learn from our past mistakes—and successes. We may even need to experiment and try on solutions. Silos have a place in farming but not in thinking.

We certainly are not the only ones who have considered that to confront our serious environmental problems, we also need to address issues such as socioeconomics and race inequality. But by suggesting that these systemic problems have a place in the environmental conversation, have we made the conversation too daunting? Is it too big?

We don't think so. We care about clean air—and having good jobs. Both can exist at the same time. We can care about worker rights and safe resource use. The communities that have been most impacted by COVID-19 are also often the ones that are most impacted by extreme weather events due to climate change.

In this book, we challenge you to examine your basic beliefs and values about the environment and do some critical and creative thinking. Where have your current ideas come from—how did they grow?

Being creative is a most helpful complement to critical thinking. By being creative, we imagine the future and the possibilities. We embrace the wonders all around us. We seek diversity in voices and perspectives.

But as Midwesterners raised on farming, the authors also believe in the need to be grounded in reality. Our families lived—and could have died—based on our success in farming. While we may find ourselves drawn to the skies in so many ways, we should not live with our heads in the clouds. With critical thinking, we are deliberate and cognitive in our search for answers. We look up, down, and all around us for solutions. These ideas may come in the form of fungi and microbes in the soil. They may come in the form of a marine animal or tropical plant that holds the cure to cancer.

In choosing to do things differently, we are accepting the possibility of uncertainty and failure. But we are also opening the door to the possibility of innovation.

In this book, we shined a light on some impediments to creative and critical thinking. There are many. But we also offer suggestions for overcoming many of these. We offer hope. These include embracing the concept of stewardship, thinking of community, and getting our hands dirty. *Planting an Idea*—or two, or three.

We also challenge readers to find inspiration in storytelling, and we share quotes from some of our favorite storytellers, such as Rachel Carson and Aldo Leopold, as well as some of our personal experiences. We suggest that critical and creative thinking can be learned. We argue that they must be learned if we are to understand nature and its value.

The child of creative and critical thinking is embracing a broad philosophical perspective that requires us to examine who we are and the role we do and must play within the natural world.

Acknowledgments

Jerry Apps

As is true of all of my books, several people have helped along the way. I especially want to thank Sam Scinta, publisher of Fulcrum Publishing, for his interest in this project and his support for my work. My wife Ruth—I call her my in-house editor—has helped me with this project as she does with all of my writing. Thank you, Ruth. I can't thank Natasha, my writing partner for this project, enough. Natasha is a fine editor, a very knowledgeable environmentalist, and a great writer.

Natasha Kassulke

The opportunity to coauthor a book with Jerry Apps is an honor. He is a Midwest treasure and beloved author and historian. Working alongside him has made me a better storyteller, and he challenged me to think critically and creatively about this project. I am grateful to my husband, Steve Apps, who shares my love of the environment and getting dirty at the family farm. Finally, I am thankful to my journalism mentor, Rob Zaleski, for his more than thirty years of support and friendship, and my former biology professor at Edgewood College, Jim Lorman, for inspiring me at a young age to look at environmental problems from various perspectives.

Notes

CHAPTER 1

1. Mindy Pennybacker, "The First Environmentalists," *The Nation*, February 7, 2000, https://www.thenation.com/article/archive/first-environmentalists/.

2. David Suzuki, *The Sacred Balance: Rediscovering Our Place in Nature* (Toronto: Greystone Books, 1997, 2002), 24.

3. EPA, "EPA History: Earth Day," https://www.epa.gov/history/epa-history-earth-day#:~: text=The%20First%20Earth%20Day%20in%20April%201970&text=Because%20 there%20was%20no%20EPA,issue%20onto%20the%20national%20agenda.

4. Erin Blakemore, "The Shocking River Fire That Fueled the Creation of the EPA," History, December 1, 2020, https://www.history.com/news/epa-earth-day-cleveland-cuyahoga-river-fire-clean-water-act.

5. Jon Hamilton, "How California's Worst Oil Spill Turned Beaches Black and the Nation Green," *Morning Edition*, NPR, January 28, 2019, https://www.npr.org/2019/01/28/688219307/how-californias-worst-oil-spill-turned-beaches-black-and-the-nation-green.

6. EPA, "EPA History: Earth Day."

7. Christine M. Whitney, "Environmental Movement," Encyclopedia.com, May 18, 2018, https://www.encyclopedia.com/earth-and-environment/ecology-and-environ-mentalism/environmental-studies/environmental-movement.

8. Tara Santora, "Earth Day to School Strikes: A Timeline of the American Environmental Movement," Stacker, April 6, 2020, https://stacker.com/stories/3968/earth-day-school-strikes-timeline-american-environmental-movement.

9. The Aldo Leopold Foundation, "Aldo Leopold," https://www.aldoleopold.org/about/aldo-leopold/.

10. The Famous People, "Frederick Law Olmsted Biography," https://www.thefamouspeople.com/profiles/frederick-law-olmsted-6671.php.

11. Daegan Miller, "A Map of Radical Bewilderment: On the Liberation Cartography of Henry David Thoreau," Places, March 2018, https://placesjournal.org/article/a-map-of-radical-bewilderment/?gclid=Cj0KCQjw6NmHBhD2ARIs

AI3hrM2IXyQm_QJrLaKwvIcm--cq1CFlAsSbT2g9G_CmX2XwV1abD6JbS8ka
Aq96EALw_wcB&cn-reloaded=1.

12. Editors of Encyclopedia Britannica, "Jens Jensen: American Landscape Architect," Britannica, https://www.britannica.com/biography/Jens-Jensen; Clearing Folk School, https://theclearing.org/wp/.

13. The John Muir Exhibit, Sierra Club, "John Muir: A Brief Biography," https://vault.sierraclub.org/john_muir_exhibit/life/muir_biography.aspx.

14. Helena Kilburn, "Leadership of Women in the Environmental Movement," Environmental Law Institute, June 12, 2019, https://www.eli.org/vibrant-environment-blog/leadership-women-environmental-movement.

15. Kilburn, "Leadership of Women."

16. Kilburn, "Leadership of Women."

17. Pennybacker, "The First Environmentalists"; Influence Watch, "Honor the Earth," https://www.influencewatch.org/non-profit/honor-the-earth/; Kilburn, "Leadership of Women."

18. Kilburn, "Leadership of Women."

19. Kilburn, "Leadership of Women."

20. NCAR/UCAR, Climate and Global Dynamics, "Warren Washington," https://www.cgd.ucar.edu/staff/wmw/.

21. SF Environment, "Celebrating Black Environmentalists during Black History Month," https://sfenvironment.org/article/celebrating-black-environmentalists-during-black-history-month.

22. Kilburn, "Leadership of Women."

23. Medium.com, "Dr. Robert Bullard: Father of Environmental Justice," https://drrobertbullard.com; Melissa Petersen, "28 Black Environmentalists, March 7, 2019, https://medium.com/@itsmelpete/black-history-month-environmentalists-69b16007da8f.

24. Editors of Encyclopedia Britannica, "Al Gore: Vice President of the United States," Britannica, March 27, 2022, www.britannica.com/biography/Al-Gore.

25. "Celebrating Black Environmentalists."

26. History, "Secretary Deb Haaland," US Department of the Interior, https://www.doi.gov/secretary-deb-haaland; "Deb Haaland, US Interior Secretary, on How She's Influenced by History," November 2, 2021, https://www.history.com/news/deb-haaland-native-american-history.

27. Editors of Encyclopedia Britannica, "Greta Thunberg: Swedish Activist," Britannica, https://www.britannica.com/biography/Greta-Thunberg.

CHAPTER 2

1. Dictionary.com, definition of "Think," https://www.dictionary.com/browse/think.
2. Cambridge Dictionary, definition of "Thinking," https://dictionary.cambridge.org/us/dictionary/english/thinking.
3. The Foundation for Critical Thinking, "Critical Thinking: Where to Begin," https://www.criticalthinking.org/pages/critical-thinking-where-to-begin/796.
4. Roger von Oech, *A Whack on the Side of the Head: How You Can Be More Creative* (New York: Business Plus [Hachette Publishing], 2008).
5. Jerry Apps, *The Quiet Season: Remembering Country Winters* (Madison: Wisconsin Historical Society Press, 2013), v.
6. Jerry Apps, *Telling Your Story: Preserve Your History through Storytelling* (Wheat Ridge, CO: Fulcrum Publishing, 2016), 8.
7. Calvin Rutstrum, *The Wilderness Life* (New York: Macmillan, 1975), 233.
8. ER Services, "Thinking Critically and Creatively," courses.lumenlearning.com/suny-foundationsacademicsuccess/chapter/thinking-critically-and-creatively/.

CHAPTER 3

1. Dictionary.com, definition of "Belief," https://www.dictionary.com/browse/belief.
2. Cambridge Dictionary, definition of "Value," https://dictionary.cambridge.org/us/dictionary/english/value.
3. University Libraries, University of Georgia, "Finding Reliable Sources: What Is a Reliable Source?" https://guides.libs.uga.edu/reliability.
4. University of Illinois–Springfield, "Freewriting," https://www.uis.edu/learning-hub/writing-resources/handouts/learning-hub/freewriting.

CHAPTER 4

1. NASA, Global Climate Change, "What's the Difference between Climate Change and Global Warming?" https://climate.nasa.gov/faq/12/whats-the-difference-between-climate-change-and-global-warming/.
2. NASA, "Climate Change: How Do We Know?" Global Climate Change, https://climate.nasa.gov/evidence/.
3. American Museum of Natural History, "How Did We Get Here?" https://www.amnh.org/exhibitions/climate-change/how-did-we-get-here.
4. Richard Black, "A Brief History of Climate Change," BBC News, September 20, 2013, https://www.bbc.com/news/science-environment-15874560.
5. AMNH, "How Did We Get Here?"

6. Sir Roland Jackson, "Who Discovered the Greenhouse Effect?" The Royal Institution, May 17, 2019, https://www.rigb.org/explore-science/explore/blog/who-discovered-greenhouse-effect.

7. Steve Graham, "John Tyndall (1820–1893)," NASA, Earth Observatory, October 8, 1999, https://earthobservatory.nasa.gov/features/Tyndall.

8. NASA, Global Climate Change, "Why Does the Temperature Record Shown on Your 'Vital Signs' Page Begin in 1880?" https://climate.nasa.gov/faq/21/why-does-the-temperature-record-shown-on-your-vital-signs-page-begin-at-1880/.

9. BBC News, "A Brief History of Climate Change."

10. BBC News, "A Brief History of Climate Change."

11. History, This Day in History, "Ford Motor Company Unveils the Model T," May 17, 2002, https://www.history.com/this-day-in-history/ford-motor-company-unveils-the-model-t.

12. Environmental Protection Agency, "Greenhouse Gases Equivalencies Calculator—Calculations and References," https://www.epa.gov/energy/greenhouse-gases-equivalencies-calculator-calculations-and-references.

13. American Chemical Society National Historic Chemical Landmarks, "The Keeling Curve: Carbon Dioxide Measurements at Mauna Loa," https://www.acs.org/content/acs/en/education/whatischemistry/landmarks/keeling-curve.html.

14. History, "Climate Change History," August 8, 2022, https://www.history.com/topics/natural-disasters-and-environment/history-of-climate-change. Thomas C. Peterson, William M. Connolley, and John Fleck, "The Myth of the 1970s Global Cooling Scientific Consensus," *Bulletin of the American Meteorological Society* 89, no. 9 (2008), DOI: 10.1175/2008BAMS2370.1.

15. History, "Climate Change History," April 7, 2022, https://www.history.com/topics/natural-disasters-and-environment/history-of-climate-change.

16. History, "Climate Change History."

17. Editors of Encyclopedia Britannica, "Kyoto Protocol: International Treaty, 1997," Britannica, https://www.britannica.com/event/Kyoto-Protocol.

18. The White House, President George W. Bush, "President Bush Discusses Global Climate Change," June 2001, https://georgewbush-whitehouse.archives.gov/news/releases/2001/06/20010611-2.html.

19. History, "Climate Change History."

20. EPA, "Global Greenhouse Gas Emissions Data," February 25, 2022, https://www.epa.gov/ghgemissions/global-greenhouse-gas-emissions-data.

21. History, "Climate Change History."

22. History, "Climate Change History."

23. Edward Wong, "Trump Has Called Climate Change a Chinese Hoax. Beijing Says Its Anything But," *New York Times*, November 19, 2016, https://www.nytimes.

com/2016/11/19/world/asia/china-trump-climate-change.html.

24. United Nations, Climate Change, "The Paris Agreement," https://unfccc.int/process-and-meetings/the-paris-agreement/the-paris-agreement.

25. History, "Climate Change History."

26. Intergovernmental Panel on Climate Change, "Summary for Policymakers of IPCC Special Report on Global Warming of 1.5°C Approved by Governments," https://www.ipcc.ch/2018/10/08/summary-for-policymakers-of-ipcc-special-report-on-global-warming-of-1-5c-approved-by-governments/.

27. United Nations, Climate Action, "2019 Climate Action Summit," https://www.un.org/en/climatechange/2019-climate-action- summit.

28. United States Census, "U.S. and World Population Clock," https://www.census.gov/popclock/.

29. Intergovernmental Panel on Climate Change, *Climate Change 2022: Impacts, Adaptation and Vulnerability—Summary for Policymakers,* February 27, 2022, https://www.ipcc.ch/report/sixth-assessment-report-working-group-ii/.

30. IPCC, *Climate Change 2022.*

31. IPCC, *Climate Change 2022.*

32. Rebecca Lindsey and LuAnn Dahlman, "Climate Change, Global Temperature," Climate.gov, April 20, 2022, https://www.climate.gov/news-features/understanding-climate/climate-change-global-temperature.

33. NASA, Global Climate Change, "The Causes of Climate Change," https://climate.nasa.gov/causes/; IPCC Sixth Assessment Report, Climate Change 2021: The Physical Science Basis, https://www.ipcc.ch/report/ar6/wg1/#SPM.

34. NOAA, PMEL Carbon Program, "What Is Ocean Acidification?" http://www.pmel.noaa.gov/co2/story/What+is+Ocean+Acidification%3F; NOAA, PMEL Carbon Program, "Ocean Acidification: The Other Carbon Dioxide Problem," https://www.pmel.noaa.gov/co2/story/Ocean+Acidification.

35. R. S. Nerem, P. D. Beckley, J. T. Fasullo, and G. T. Mitchum, "Climate- Change-Driven Accelerated Sea-Level Rise Detected in the Altimeter Era," PNAS, February 12, 2018, https://www.pnas.org/doi/abs/10.1073/pnas.1717312115.

36. J. Hatfield, G. Takle, R. Grotjahn, P. Holden, R. C. Izaurralde, T. Mader, E. Marshall, and D. Liverman, in Chapter 6, "Agriculture," *Climate Change Impacts in the United States: The Third National Climate Assessment,* ed. J. M. Melillo, Terese (T. C.) Richmond, and G. W. Yohe (U.S. Global Change Research Program, 2014), 150–174, https://nca2014.globalchange.gov/report/sectors/agriculture.

37. S. Pryor, C., D. Scavia, C. Downer, M. Gaden, L. Iverson, R. Nordstrom, J. Patz, and G. P. Robertson, in Chapter 18, "Midwest." *Climate Change Impacts in the United States: The Third National Climate Assessment,* ed. J. M. Melillo, Terese (T.C.) Richmond, and

G. W. Yohe, (U.S. Global Change Research Program, 2014), 418–440.

38. EPA, Climate Change Impacts, "Climate Impacts in the Midwest," https://climatechange.chicago.gov/climate-impacts/climate-impacts-midwest.

39. NASA, "Causes of Climate Change."

40. Jason Samenow, "Death Valley Soars to 130 Degrees, Matching Earth's Highest Temperature in at Least 90 Years," *Washington Post*, July 9, 2021, https://www.washingtonpost.com/weather/2021/07/09/death-valley-record-high-temperature/.

41. Matthew Cappucci and Jason Samenow, "Heat Wave Blasts U.S. with 150 Million Americans Under Alerts," *Washington Post*, August 12, 2021, https://www.washingtonpost.com/weather/2021/08/11/heatwave-united-states-pacific-northwest/.

42. Jan Wesner Childs, "July Was Earth's Hottest Month on Record," The Weather Channel, August 13, 2021, https://weather.com/science/environment/news/2021-08-13-hottest-month-on-earth-july.

43. NOAA, "June 2021 Was the Hottest June on Record for U.S.," July 9, 2021, https://www.noaa.gov/news/june-2021-was-hottest-june-on-record-for-us.

44. David Williams, "Extreme Heat Cooked Mussels, Clams and Other Shellfish Alive on Beaches in Western Canada," CNN, July 12, 2021, https://www.cnn.com/2021/07/10/weather/heat-sea-life-deaths-trnd-scn/index.html.

45. NOAA, "Climate Data Online," https://www.ncdc.noaa.gov/cdo-web/.

46. NOAA, "Climate Data Online."

47. Audubon, "New Audubon Science: Two-Thirds of American Birds at Risk of Extinction Due to Climate Change," October 10, 2019, https://ca.audubon.org/news/new-audubon-science-two-thirds-north-american-birds-risk-extinction-due-climate-change.

48. The Nature Conservancy, "The Language of Climate and Clean Energy," July 18, 2018, https://alliancerally.org/wp-content/ uploads/2018/05/Rally2018_C05-Language-of-Climate-and-Clean-Energy.pdf.

49. Justin Worland, "Force of Nature: The Pandemic Remade the Economy. Now, It's the Climate's Turn," *Time,* April 26/May 3, 2021, 60–66.

50. Cook et al., "Consensus on Consensus: A Synthesis of Consensus Estimates on Human-Caused Global Warming," *Environmental Research Letters* 11, no. 4 (April 13, 2016), DOI:10.1088/1748-9326/11/4/048002.

51. AAAS.org Staff Report, "AAAS Reaffirms Statements on Climate Change and Integrity," American Association for the Advancement of Science, December 4, 2009, https://www.aaas.org/news/aaas-reaffirms-statements-climate-change-and-integrity.

52. California Governor's Office of Planning and Research, "List of Worldwide Scientific Organizations," https://www.opr.ca.gov/facts/list-of-scientific-organizations.html.

53. O. Edenhofer, R. Pichs-Madruga, Y. Sokona, et al., eds., "Summary for Policymakers"

in *Climate Change 2014: Mitigation of Climate Change. Contribution of Working Group III to the Fifth Assessment Report of the Intergovernmental Panel on Climate Change* (Cambridge, UK, and New York: Cambridge University Press, 2014), https://www.ipcc.ch/site/assets/uploads/2018/02/ipcc_wg3_ar5_summary-for-policymakers.pdf.

54. Cary Funk and Meg Hefferon, "U.S. Public Views on Climate and Energy," Pew Research Center, November 25, 2019, https://www.pewresearch.org/science/2019/11/25/u-s-public-views-on-climate-and-energy/.

55. Olafur Eliasson and Minik Thorleif Rosing, "Ice Watch," https://icewatch.london.

56. U.S. Energy Information and Administration, "Annual Coal Report," October 4, 2021, https://www.eia.gov/coal/annual/.

57. Funk and Hefferon, "U.S. Public Views on Climate Change and Energy."

58. Dean Scott, "UN Climate Report's Warnings Compound Worries in Insurance World," Bloomberg Law, August 16, 2021, https://newsbloomberglaw.com/environment-and-energy/un-climate-reports-warnings-compound-worries-in-insurance-world.

59. Centers for Disease Control and Prevention Newsroom, "Illnesses from Mosquito, Tick, and Flea Bites Increasing in the US," May 1, 2018, https://www.cdc.gov/media/releases/2018/p0501-vs-vector-borne.html.

60. Girija Syamlal, Brent Doney, and Jacek M. Mazurek, "Chronic Obstructive Pulmonary Disease Prevalence Among Adults Who Have Never Smoked, by Industry and Occupation—United States, 2013–2017," Centers for Disease Control and Prevention, April 5, 2019, https://www.cdc.gov/mmwr/volumes/68/wr/mm6813a2.htm.

61. Simon Hattenstone, "The Transformation of Greta Thunberg," *The Guardian*, September 25, 2021, https://www.theguardian.com/environment/ng-interactive/2021/sep/25/greta-thunberg-i-really-see-the-value-of-friendship-apart-from-the-climate-almost-nothing-else-matters.

62. Greta Thunberg, Twitter, July 7, 2021, https://twitter.com/gretathunberg/status/1412843637982076931; Peter Gleick, "The Climate Crisis Will Create Two Classes: Those Who Can Flee, and Those Who Cannot," *The Guardian*, July 7, 2021, https://www.theguardian.com/commentisfree/2021/jul/07/global-heating-climate-crisis-heat-two-classes?CMP=Share_AndroidApp_Other.

63. Terry Devitt, "No Snow, No Hares: Climate Change Pushes Emblematic Species North," University of Wisconsin–Madison News, March 30, 2016, https://news.wisc.edu/no-snow-no-hares-climate-change-pushes-emblematic-species-north/.

64. NASA, Global Climate Change, "A Degree of Concern: Why Global Temperatures Matter," June 19, 2019, https://news.wisc.edu/no-snow-no-hares-climate-change-pushes-emblematic-species-north/.

65. Ciara Nuget, "The Enduring Hope of Jane Goodall," *Time*, October 11/October 18, 2021, ISSN 0040-781X, 30–41, https://time.com/6102640/jane-goodall-environment-hope/.

CHAPTER 5

1. USDA, Economic Research Service, "Farming and Farm Income," February 4, 2022, https://www.ers.usda.gov/data-products/ag-and-food-statistics-charting-the-essentials/farming-and-farm-income/.

2. Find My Past, US Census 1830, https://www.findmypast.com/articles/world-records/full-list-of-united-states-records/census-land-and-substitutes/us-census-1830.

3. Jeff Hoyt, "1800–1990: Changes in Urban/Rural U.S. Population," Seniorliving.org, June 29, 2021, https://www.seniorliving.org/history/1800-1990-changes-urbanrural-us-population/; 1940 Census, "About the 1940 Census," https://1940census.archives.gov/about.asp?

4. Brandon McBride, "Celebrating the 80th Anniversary of the Rural Electrification Administration," USDA, February 21, 2017, https://www.usda.gov/media/blog/2016/05/20/celebrating-80th-anniversary-rural-electrification-administration.

5. USDA, Economic Research Service, "The Number of U.S. Farms Continues to Decline Slowly," May 10, 2021, https://www.ers.usda.gov/data-products/chart-gallery/gallery/chart-detail/?chartId=58268.

6. Carolyn Dimitri, Anne Effland, and Neilson Conklin, "The 20th Century Transformation of U.S. Agriculture and Farm Policy," USDA, Economic Research Service, Economic Information Bulletin Number 3, https://www.ers.usda.gov/webdocs/publications/44197/13566_eib3_1_.pdf.

7. Sara Popescu Slavikova, "Advantages and Disadvantages of Monoculture Farming," greentumble.com, June 16, 2019, https://greentumble.com/advantages-and-disadvantages-of-monoculture-farming/.

8. Slavikova, "Advantages and Disadvantages of Monoculture Farming."

9. Chris Hubbuck, "DNR Study Finds Irrigation Responsible for Draining Lakes," *Wisconsin State Journal*, April 6, 2021.

10. Eric Hamilton, "Midwest Bumble Bees Declined with More Farmed Land, Less Diverse Crops Since 1870," UW–Madison News Release, June 22, 2021, https://news.wisc.edu/midwest-bumble-bees-declined-with-more-farmed-land-less-diverse-crops-since-1870/.

11. P. Byrne, "Genetically Modified (GM) Crops: Techniques and Applications," Colorado State University Extension, https://extension.colostate.edu/docs/pubs/crops/00710.pdf.

12. Sheldon Krimsky and Jeremy Gruber, eds., *The GMO Deception* (New York: Skyhorse Publishing, 2014), xxi–xxv.

13. Union of Concerned Scientists, "What Is Sustainable Agriculture"? March 15, 2022, https://www.ucsusa.org/resources/what-sustainable-agriculture?gclid=CjwKCAjwt8uGBhBAEiwAayu_9UKDgjUbWX0RG0XmgdhhhtKAvzrNljZx3tYMyuYUQUtcj6r5g94iaRoC.

14. Raoul Adamchak, "Organic Farming," Britannica, https://www.britannica.com/topic/organic-farming.

15. N.A., *Waushara Argus*—Farm Section, June 24, 2021.

16. Pinduoduo, "The Pros and Cons of Aquaculture," April 20, 2021, https://stories.pinduoduo-global.com/agritech-hub/pros-and-cons-of-aquaculture.

17. Eden Green Technology, "How Hydroponics Is Revolutionizing Farming Jobs in the Agriculture Industry," March 25, 2021, https://www.edengreen.com/blog-collection/how-hydroponics-is-revolutionizing-farming-jobs.

18. Hannah Rodriguez, "A Farm Without Fields," *Wisconsin State Farmer*, July 23, 2021, 5B–6B.

19. Rob Dongoski and EYAmericas, "Mark Oshima—Farming Up," EY Americas, July 15, 2020, https://www.ey.com/en_us/purpose/next-up-mark-oshima-farming-up.

20. Clean Water Action Council of Northeast Wisconsin, "Factory Farming Impacts," https://www.cleanwateractioncouncil.org/issues/resource-issues/factory-farms/.

21. Wisconsin Department of Natural Resources, "Concentrated Animal Feeding Operations," https://dnr.wisconsin.gov/topic/cafo; USDA, National Agricultural Statistics Service, "2011 Wisconsin Agricultural Statistics," August 2011, https://www.nass.usda.gov/Statistics_by_State/Wisconsin/Publications/Annual_Statistical_Bulletin/bulletin2011_web.pdf.

22. Livestock and Poultry Environmental Learning Community, "Liquid Manure Storage Ponds, Pits, and Tanks," March 5, 2019, https://lpelc.org/liquid-manure-storage-ponds-pits-and-tanks/; Rick Barrett and Lee Bergquist, "Industrial Dairy Farming Is Taking Over in Wisconsin, Crowding Out Family Operations and Raising Environmental Concerns," *Milwaukee Journal Sentinel*, December 6, 2019, https://www.jsonline.com/in-depth/news/special-reports/dairy-crisis/2019/12/06/industrial-dairy-impacts-wisconsin-environment-family-farms/4318671002/.

23. Sarah Whites-Koditschek and Coburn Dukehart, "New Research Indicates Tainted Kewaunee County Wells Tied to Manure Pits," *Green Bay Press Gazette*, March 4, 2019, https://www.greenbaypressgazette.com/story/news/investigations/2019/03/04/tainted-kewaunee-county-drinking-water-wells-tied-manure-pits/3054018002/; Chris Rickert, "Study of Southwest Wisconsin Well Water Continues to Indicate Contamination," *Wisconsin State Journal*, April 19, 2020, https://madison.com/wsj/news/local/environment/study-of-southwest-wisconsin-well-water-continues-to-indicate-contamination/article_673d85cd-c383-5b7e-93b9-892f54ca6091.html.

24. Will Cushman, "CAFO Oversight in Wisconsin and Who Pays for It?" *Wisconsin State Farmer*, June 19, 2019, https://www.wisfarmer.com/story/news/2019/06/19/what-cafo-oversight-and-who-pays-wisconsin/1452486001/.

25. A Greener World, "Pollution: U.S. Industrial Livestock Farms Produce Up to 1.37 Billion Tons of Manure Each Year—That's 20 Times More Fecal Waste Than the Entire U.S. Human Population, https://agreenerworld.org/challenges-and-opportunities/environmental-pollution/.

26. Coburn Dukehart "Cow Manure Predicted to Cause Most Sickness from Contaminated Wells in Kewaunee County, *Wisconsin Watch*, June 28, 2021, https://wisconsinwatch.org/2021/06/cow-manure-predicted-to-cause-most-sickness-from-contaminated-wells-in-kewaunee-county/.

27. Aldo Leopold, "The Land Ethic," in Beth Waterhouse, PBS Online: "Death of the Dream: A Sustainable Future," https://www.pbs.org/ktca/farmhouses/sustainable.html.

28. Wendell Berry, "The Agrarian Standard," in *The Essential Agrarian Reader: The Future of Culture, Community, and the Land*, ed. Norman Wirzba (Lexington: The University Press of Kentucky, 2003), 24–26.

29. EPA, "Basic Information about Anaerobic Digestion (AD)," https://www.epa.gov/anaerobic-digestion/basic-information-about-anaerobic-digestion-ad.

30. Sierra Club, "Sierra Club Guidance: Methane Digesters and Concentrated Animal Feeding Operation (CAFO) Waste," https://www.sierraclub.org/sites/www.sierraclub.org/files/methane_digesters.pdf.

31. Stefani Sassos, "All of the Nutritional Facts and Health Benefits of Goat Milk," *Good Housekeeping*, April 14, 2020, https://www.goodhousekeeping.com/health/diet-nutrition/a32068757/goat-milk-health-benefits/.

32. Pam Jahnke and Bob Bosold, "Farmers for Sustainable Food," The Mid-West Farm Report, May 11, 2021, https://www.midwestfarmreport.com/2021/05/11/farmers-for-sustainable-food/.

33. USDA, Economic Research Service, "Dairy Products: Per Capita Consumption, United States," September 4, 2020, https://www.ers.usda.gov/data-products/dairy-data/.

34. Susan S. Lang, "Lactose Intolerance Seems Linked to Ancestral Struggles with Harsh Climate and Cattle Diseases, Cornell Study Finds," *Cornell Chronicle*, June 1, 2005, https://news.cornell.edu/stories/2005/06/lactose-intolerance-linked-ancestral-struggles-climate-diseases.

35. Perfect Day, https://perfectday.com.

36. Alexandra Wilson, "Got Milk? This $40M Startup Is Creating Cow-Free Dairy Products That Actually Taste Like the Real Thing," *Forbes*, January 9, 2019, https://www.forbes.com/sites/alexandrawilson1/2019/01/09/got-milk-this-40m-startup-is-creating-cow-free-dairy-products-that-actually-taste-like-the-real-thing/?sh=7005c96240dc.

CHAPTER 6

1. Dede Mulligan, "The History of Logging in the USA," Premier Firewood Co., April 5, 2016, https://www.premierfirewoodcompany.com/blog/2016/04/05/the-history-of-logging-in-the-usa/#:~:text=The%20logging%20industry%20began%20in,supplying%20lumber%20throughout%20the%20world.

2. Society of American Foresters, "History," https://www.premierfirewoodcompany.com/blog/2016/04/05/the-history-of-logging-in-the-usa/.

3. USDA Forest Service, "Our History," https://www.fs.usda.gov/learn/our-history.

4. Jerry Apps, *When the White Pine Was King* (Madison: Wisconsin Historical Society Press, 2020).

5. Increase Allen Lapham, Joseph Gillett Knapp, and H. Crocker, *Report on the Disastrous Effect on the Destruction of Forest Trees Now Going on Rapidly in the State of Wisconsin* (Madison, WI: Atwood and Rublee, State Printer, 1867), 100.

6. Frederick Merk, *Economic History of Wisconsin During the Civil War Decade* (Madison: Wisconsin Historical Society Press, 1916), 110.

7. Wisconsin County Forests Association, Forestry Division, DNR, "Frequently Asked Questions about WI Forests," https://wisconsincountyforests.com.

8. Center for Disaster Philanthropy, "2020 North American Fire Season," https://disasterphilanthropy.org/disasters/2020-california-wildfires/.

9. EPA, "Climate Impacts on Forests," https://19january2017snapshot.epa.gov/climate-impacts/climate-impacts-forests_.html.

10. Chris Deziel, "Environmental Problems Caused by Deforestation of Tropical Rainforests," Sciencing, July 16, 2018, https://sciencing.com/environmental-problems-caused-deforestation-tropical-rain-forests-22487.html.

11. Joe Hovel, *Northwoods Forest Conservation: Managing Forestlands for the Future* (Conover, WI: Partners in Forestry Cooperative and Northwoods Alliance, 2021), 17.

12. Hovel, *Northwoods Forest Conservation*, 3–7.

13. Michael Jacobson and Sanford S. Smith, "Sustainable Forestry," PennState Extension, September 8, 2016, https://extension.psu.edu/sustainable-forestry.

14. USDA Forest Service, "Forest Legacy," https://www.fs.usda.gov/managing-land/private-land/forest-legacy.

15. Wisconsin Department of Natural Resources, "Managed Forest Law," https://dnr.wisconsin.gov/topic/forestlandowners/mfl.

CHAPTER 7

1. Kimberly Mullen, "Information on Earth's Water," NGWA, https://www.ngwa.org/what-is-groundwater/About-groundwater/information-on-earths-water.

2. Rima Hanano, "Save Water: Reduce Your Water Footprint," Reset—Digital for Good, March 3, 2010, https://en.reset.org/save-water-reduce-your-water-footprint/.

3. "Why We Need Water for the Environment," NSW Environment, Energy and Science, https://www.google.com/search?q=Why+We+Need+Water+for+the+Environment% 2C%E2%80%9D+NSW+Environment%2C+Energy+and+Science&sxsrf=ALiCzsbs WGISbEPSTgnv7Gjw5aefVlafgg%3A1659107801458&source=hp&ei=2fnjYvu GGe6U0PEP786-0Ag&iflsig=AJiK0e8AAAAAYuQH6a2qzXQnSMN3IJagrA.

4. Paul Srubas, "As PCB Cleanup Nears End, Area Mills Appear to Have Agreed on Payment," *Green Bay Press Gazette*, November 3, 2017, https://www.greenbaypressgazette.com/story/news/2017/11/03/fox-cleanup-legal-pcb/817625001/.

5. IUCN, "Marine Plastic Pollution," November 2021, https://www.iucn.org/resources/issues-briefs/marine-plastic-pollution.

6. Melissa Denchak, "Water Pollution: Everything You Need to Know," NRDC, April 18, 2022, https://www.nrdc.org/stories/water-pollution-everything-you-need-know#types.

7. "Summary of the Clean Water Act, US EPA, https://www.epa.gov/laws-regulations/summary-clean-water-act.

8. Denchak, "Water Pollution."

9. World Wildlife Fund, "Water Scarcity," Threats, https://www.worldwildlife.org/threats/water-scarcity.

10. Julia Lurie, "California's Almonds Suck as Much Water Annually as Los Angeles Uses in Three Years," *Mother Jones*, January 12, 2015, https://www.motherjones.com/environment/2015/01/almonds-nuts-crazy-stats-charts/.

11. Food and Water Watch, "Take the Pledge—Take Back the Tap," https://secure.foodandwaterwatch.org/act/take-pledge-take-back-tap?ms=onad_gg_2018000_ Take-Back-Tap&oms=onad_gg_2018000_Take-Back-Tap&gclid=CjwKCAjwg b6IBhAREiwAgMYKRiKqZXNRnJmjZ9Z1xWdzEeJnw_M7IYrhkQ2SU2fjo0XT V9mmsZ2XuRoC960QAvD_BwE.

12. Wisconsin Academy of Sciences, Arts and Letters, *Waters of Wisconsin: The Future of Our Aquatic Ecosystems and Resources* (Madison: Wisconsin Academy of Science, Arts and Letters, 2003).

13. The National Agricultural Law Center, "Water Law: An Overview," https://nationalaglawcenter.org/overview/water-law/.

CHAPTER 8

1. National Rural Electric Cooperative Association, "The Electric Cooperative Story," https://www.electric.coop/ourorganization/history#:~:text=On%20May%20 11%2C%201935%2C%20Roosevelt,became%20the%20REA%20got%20underway.

2. Leslie C. Mcmanus, "Charged by the Wind," Farm Collector, https://www.farm collector.com/equipment/charged-by-the-wind/.

3. U.S. Energy Information Administration, "History of Energy Consumption in the United States, 1775–2009," https://www.farmcollector.com/equipment/charged-by-the-wind/.

4. Melissa Denchak, "Fracking 101," NRDC, April 19, 2019, https://www.nrdc.org/stories/fracking-101.

5. ProCon.org, "Fracking—Top 3 Pros and Cons," January 21, 2022, https://www.procon.org/headlines/fracking-top-3-pros-and-cons/.

6. Wisconsin Department of Natural Resources, "Industrial Sand Mining Overview," https://dnr.wisconsin.gov/topic/Mines/Sand.html.

7. Dr. Thomas Power and Donovan S. Power, "The Economic Benefits and Costs of Frac-Sand Mining in West Central Wisconsin," Institute for Agriculture and Trade Policy, May 15, 2013, https://www.iatp.org/frac-report.

8. US Energy Information Association, "Renewable Energy Explained," May 20, 2021, https://www.eia.gov/energyexplained/renewable-sources/.

9. US Energy Information Association, "Electricity Explained," April 19, 2022, https://www.eia.gov/energyexplained/electricity/electricity-in-the-us.php.

10. US Energy Information Association, "Use of Energy Explained," May 14, 2021, https://www.eia.gov/energyexplained/use-of-energy/.

11. EIA, "Use of Energy Explained."

12. US Energy Information Association, "Hydropower Explained," March 16, 2022, https://www.eia.gov/energyexplained/hydropower/.

13. U.S. Energy Information Association, "Geothermal Explained," February 15, 2022, https://www.eia.gov/energyexplained/geothermal/where-geothermal-energy-is-found.php.

14. USDA Economic Research Service, "Feedgrains Sector at a Glance," June 28, 2021, https://www.ers.usda.gov/topics/crops/corn-and-other-feedgrains/feedgrains-sector-at-a-glance/.

15. Danny Ovy, "Advantages and Disadvantages of Ethanol," Alternative-Energies.net, February 10, 2018, https://www.alternative-energies.net/advantages-and-disadvantages-of-ethanol/.

16. Energy.gov, "Planning for Home Renewable Energy Systems," https://www.energy.gov/energysaver/planning-home-renewable-energy-systems.

17. Volt, "Why Choose LED Lighting?" https://www.voltlighting.com/learn/why-choose-led-lighting?gclid=CjwKCAjwgviIBhBkEiwA10D2j9Ryn5m_f-jeI9IqceX4-kbTKyG555tEAUV_QIdWY7JqAnqOpyveXRoC4fMQAvD_BwE.

18. Save on Energy, "12 Ways to Make Your Home More Energy Efficient," https://saveonenergy.ca/For-Your-Home/Advice-and-Tips/12-ways-to-make-your-home-more-energy-efficient; U.S. EPA, "Reduce the Environmental Impact of Your Energy Use," https://www.epa.gov/energy/reduce-environmental-impact-your-energy-use; and Dawn Jamison, "16 Ways to Make Your Home More Energy-Efficient," Zing, by Quicken Loans, March 11, 2016, https://www.quickenloans.com/blog/16-ways-to-make-your-home-more-energy-efficient.

19. Benefits.gov, Weatherization Assistance Program for Low-Income Persons, https://www.benefits.gov/benefit/580.

CHAPTER 9

1. Environmental Protection, "Wisconsin DNR Reminds Residents of Trash Burning Rules," April 6, 2011, https://townofcalumet.com/wp-content/uploads/2020/02/110406-Wisconsin-DNR-Reminds-Residents-of-Trash-Burning-Rules-Environmental-Protection.pdf.

2. EPA, "Evolution of the Clean Air Act," December 7, 2021, https://www.epa.gov/clean-air-act-overview/evolution-clean-air-act.

3. Stephen Leahy, "Without the Ozone Treaty You'd Get Sunburned in 5 Minutes," *National Geographic*, September 24, 2017, https://www.nationalgeographic.com/science/article/montreal-protocol-ozone-treaty-30-climate-change-hcfs-hfcs.

4. Leahy, "Without the Ozone Layer You'd Get Sunburned in 5 Minutes."

5. Christopher S. Malley, Daven K. Henze, Johan C.I. Kuylenstierna, et al., "Updated Global Estimates of Respiratory Mortality in Adults ≥30Years of Age Attributable to Long-Term Ozone Exposure," Environmental Health Perspectives, August 28, 2017, https://ehp.niehs.nih.gov/doi/10.1289/ehp1390.

6. EPA, "Health and Environmental Effects of Particulate Matter (PM)," May 26, 2021, https://www.epa.gov/pm-pollution/health-and-environmental-effects-particulate-matter-pm.

7. Andrew Freedman, "Air Pollution Shaves Off 2.2 Years of Average Life Expectancy Worldwide," Axios, September 1, 2021, https://www.axios.com/air-pollution-global-life-expectancy-report-fb6a821c-95aa-4671-8005-caac8be0e69b.html.

8. Jim Robbins, "Ozone Pollution: An Insidious and Growing Threat to Biodiversity," Yale Environment 360, October 7, 2021, https://e360.yale.edu/features/ozone-pollution-an-insidious-and-growing-threat-to-biodiversity.

9. Robbins, "Ozone Pollution."

10. Robbins, "Ozone Pollution."

11. Amos P. K. Tai and Maria Val Martin, "Impacts of Ozone Air Pollution and Temperature Extremes on Crop Yields: Spatial Variability, Adaptation and Implications

for Future Food Security," Science Direct, November 2017, https://www.sciencedirect.com/science/article/pii/S1352231017305836.

12. N. E. Grulke, R. A. Minnich, T. Paine, and P. Riggan, "Air Pollution Increases Forest Susceptibility to Wildfires: A Case Study for the San Bernardino Mountains in Southern California," USDA Forest Service, 2010, https://www.fs.usda.gov/pnw/publications/air-pollution-increases-forest-susceptibility-wildfires-case-study-san-bernardino-0.

13. National Interagency Fire Center, "Fire Information," https://www.nifc.gov/fire-information.

14. National Park Service, "Wildfire Causes and Evaluations," March 8, 2022, https://www.nps.gov/articles/wildfire-causes-and-evaluation.htm.

15. L. Shi, P. Liu, A. Zanobetti, and J. Schwartz, "Climate Penalty: Climate-Driven Increases in Ozone and PM2.5 Levels and Mortality" (abstract), *Environmental Epidemiology*, 3 (October 2019): 35, https://journals.lww.com/environepidem/Fulltext/2019/10001/Climate_Penalty__Climate_driven_increases_in_ozone.1116.aspx.

16. EPA, "What Is Acid Rain?" December 3, 2021, https://www.epa.gov/acidrain/what-acid-rain.

17. Water Science School, "Acid Rain and Water," USGS, August 2, 2019, https://www.usgs.gov/special-topics/water-science-school/science/acid-rain-and-water?qt-science_center_objects=0#qt-science_center_objects.

18. National Science Foundation, "Acid Rain: Scourge of the Past or Trend of the Present?" July 25, 2012, https://beta.nsf.gov/news/acid-rain-scourge-past-or-trend-present.

19. EPA, "Benefits and Costs of the Clean Air Act 1990–2020, the Second Prospective Study," August 12, 2021, https://www.epa.gov/clean-air-act-overview/benefits-and-costs-clean-air-act-1990-2020-second-prospective-study.

20. EPA, "Protecting Our Nation's Treasured Vistas," https://epa.maps.arcgis.com/apps/Cascade/index.html?appid=e4dbe2263e1f49fb849af1c73a04e2f2.

21. EPA, "National Air Quality: Status and Trends of Key Air Pollutants," June 1, 2020, https://www.epa.gov/air-trends.

22. EPA, "History of Reducing Air Pollution from Transportation in the United States," May 16, 2022, https://www.epa.gov/transportation-air-pollution-and-climate-change/accomplishments-and-success-air-pollution-transportation.

23. EPA, "Regulations for Emissions from Locomotives," May 24, 2022, https://www.epa.gov/regulations-emissions-vehicles-and-engines/regulations-emissions-locomotives.

24. EPA, "Stationary Sources of Air Pollution, August 12, 2021, https://www.epa.gov/stationary-sources-air-pollution.

25. EPA, "New Interactive Maps and Resources Empower the Public and Policymakers to Act on Environmental Justice," July 29, 2021, https://www.epa.gov/newsreleases/new-interactive-maps-and-resources-empower-public-and-policymakers-

act-environmental.

26. EPA, "New Interactive Maps and Resources."

27. Rainforest Rescue, "Palm Oil—Deforestation for Everyday Products," https://www.rainforest-rescue.org/topics/palm-oil#start.

28. EPA, "Actions You Can Take to Reduce Air Pollution," https://www3.epa.gov/region1/airquality/reducepollution.html.

29. Isaac Asimov, *I, Asimov: A Memoir* (New York, Bantam Books, 1995).

CHAPTER 10

1. Larry Nesper, *The Struggle for Ojibwe Spearfishing and Treaty Rights* (Lincoln: University of Nebraska Press, 2002).

2. Erin Ryan, "A Short History of the Public Trust Doctrine and Its Intersection with Private Water Law," *SSRN Electronic Journal* 39 (December 2020), DOI: 10.2139/ssrn.3648637, https://www.researchgate.net/publication/342851691_A_Short_History_of_the_Public_Trust_Doctrine_and_its_Intersection_with_Private_Water_Law.

3. United Nations Peacekeeping, "Conflict and Natural Resources," https://peacekeeping.un.org/en/conflict-and-natural-resources.

4. Millennium Alliance for Humanity and the Biosphere Admin., "When Fossil Fuels Run Out, What Then?" MAHB, May 23, 2019, https://mahb.stanford.edu/library-item/fossil-fuels-run/.

5. Marie-Catherine Reikhof, Esther Regnier, and Martin F. Quass, "Economic Growth, International Trade, and the Depletion or Conservation of Renewable Natural Resources," *Journal of Environmental Economics and Management*, 97 (September 2019), https://www.sciencedirect.com/science/article/abs/pii/S0095069616303254?via%3Dihub.

6. United Nations, Department of Economic and Social Affairs, "World Population Projected to Reach 9.8 Billion in 2050, and 11.2 Billion in 2100," June 21, 2017, https://www.un.org/development/desa/en/news/population/world-population-prospects-2017.html.

7. Oren Lyons, "An Iroquois Perspective," in *American Indian Environments: Ecological Issues in Native American History*, ed. Christopher Vecsey and Robert W. Venables (New York: Syracuse University Press, 1980), 173–174.

8. Wisconsin Department of Natural Resources, "Knowles-Nelson Stewardship Program," https://dnr.wisconsin.gov/topic/Stewardship.

CHAPTER 11

1. Winona LaDuke, *All My Relations: Native Struggles for Land and Life* (Cambridge, MA: South End Press, 1999), 2.

2. Jerry Apps, *Whispers and Shadows: A Naturalist's Memoir* (Madison: Wisconsin Historical Society Press, 2015), xiv.

3. Jack D. Forbes, "Indigenous Americans: Spirituality and Ecos," *Daedalus* (Fall 2001), https://www.amacad.org/publication/indigenous-americans-spirituality-and-ecos.

4. Aldo Leopold, *A Sand County Almanac* (New York: Oxford University Press, 1949), Foreword.

5. Apps, *Whispers and Shadows*, xiii–xiv.

6. USDA, Economic Research Service, "Major Land Uses," October 20, 2019, https://www.ers.usda.gov/topics/farm-economy/land-use-land-value-tenure/major-land-uses/; see also https://www.ers.usda.gov/data-products/major-land-uses/.

7. USDA, "Major Land Uses."

8. Food and Agriculture Organization of the United Nations, "The State of the World's Forests 2020," https://www.fao.org/state-of-forests/en/.

9. Christina Nunez, "Deforestation Explained," *National Geographic*, https://www.nationalgeographic.com/environment/article/deforestation.

10. EPA, "Basic Information about Nonpoint Source (NPS) Pollution," July 8, 2021, https://www.epa.gov/nps/basic-information-about-nonpoint-source-nps-pollution.

11. EPA, "Basic Information about Nonpoint Source (NPS) Pollution."

12. EPA, "About Smart Growth," March 28, 2022, https://www.epa.gov/smartgrowth/about-smart-growth.

13. As quoted in A. L. Booth, "We Are the Land: Native American Views of Nature," in *Nature Across Cultures: Science Across Cultures: The History of Non-Western Science*, vol. 4, ed. H. Selin (Dordrecht, Netherlands: Springer, 2003), https://link.springer.com/chapter/10.1007/978-94-017-0149-5_17.

14. Apps, *Whispers and Shadows*, 70.

15. Apps, *Whispers and Shadows*, 127.

16. LSPIRG, "Know the Land Territories Campaign," http://www.lspirg.org/knowtheland.

17. Wisconsin Geological and Natural History Survey, "Ice Age Geology," https://wgnhs.wisc.edu/wisconsin-geology/ice-age/.

18. Craig Mattson, "The Importance of Glaciers to Wisconsin," Schlitz Audubon Nature Center, https://www.schlitzaudubon.org/2019/01/15/the-importance-of-glaciers-to-wisconsin/.

CHAPTER 12

1. Jill Utrup and Chris Mensing, "*Lycaeides melissa samuelis*," US Fish and Wildlife Service, March 11, 2019, https://www.fws.gov/species/karner-melissa-blue-lycaeides-melissa-samuelis.

2. Maryann Mott, "U.S. Endangered Species Act Works, Study Finds," National Geographic

News, https://www.biologicaldiversity.org/news/media-archive/US%20Endangered %20Species%20Act%20Works.pdf.

3. US Fish and Wildlife Service, Environmental Conservation Online System, "Listed Species Summary (Boxscore)," https://ecos.fws.gov/ecp/report/boxscore.

4. Hannah Ritchie and Max Roser, "Extinctions," Our World in Data, 2021, https://ourworldindata.org/extinctions.

5. A. D. Barnosky, N. Matzke, S. Tomiya, et al., "Has the Earth's Sixth Mass Extinction Already Arrived?" Review, *Nature*, 471 (March 3, 2011), https://www.nature.com/articles/nature09678.epdf?sharing_token=MgXie0x6iEFOQbbti92NBdRgN0jAjWel-9jnR3ZoTv0Naoo3AqTwA8fP2JzR6E5I3QXGbc_Emc8jHO9Wm4938FZxuXvcyt-P6VnDCaOZdDQyMRfkQzOe1T5Fwg1NKWUlmXWQhRTu2xtoMOxB7KrA-rPna-bovdq-0XgeRP7SKnnbJETpnludKLF8JWeeQ3a4ck0.

6. M. J. Novacek, ed., *The Biodiversity Crisis: Losing What Counts* (New York: The New Press, 2001).

7. Ritchie and Roser, "Extinctions," https://ourworldindata.org/extinctions#citation.

8. Henry Mall, "As Milkweed Goes, So Goes the Monarch," *Grow*, Fall 2020, https://grow.cals.wisc.edu/deprecated/on-henry-mall/as-milkweed-goes-so-goes-the-monarch.

9. IUCN, "Migratory Monarch Butterfly Now Endangered—IUCN Red List," July 21, 2022, https://www.iucn.org/press-release/202207/migratory-monarch-butterfly-now-endangered-iucn-red-list.

10. D. B. Wake and V. T. Vredenburg, *Are We in the Midst of the Sixth Mass Extinction? A View from the World of Amphibians*, August 12, 2008, 105: 11466–11473, http://www.pnas.org/content/early/2008/08/08/0801921105.abstract.

11. Alex Daniel, "The Science-Backed Reasons Why Saving Endangered Species Is Important," February 27, 2019, https://bestlifeonline.com/saving-endangered-species/.https://www.fws.gov/project/critical-habitat.

12. Ritchie and Roser, "Extinctions."

13. U.S. Fish & Wildlife Service, "Critical Habitat," https://www.fws.gov/project/critical-habitat.

14. Nathan Rott, "Once Imperiled, America's Bald Eagle Populations Are Soaring," NPR, CPR News, March 25, 2021, https://www.npr.org/2021/03/25/981272794/once-imperiled-americas-bald-eagle-populations-are-soaring.

15. The National Wildlife Federation, "Northern Spotted Owl," https://www.nwf.org/Educational-Resources/Wildlife-Guide/Birds/Northern-Spotted-Owl.

16. William Yardley, "Plan Issued to Save Northern Spotted Owl," June 30, 2011, *New York Times*, https://www.nytimes.com/2011/07/01/us/01owls.html.

17. USGS, "Northern Spotted Owl Still Fights for Survival," October 6, 2021, https://

www.usgs.gov/news/featured-story/northern-spotted-owl-still-fights-survival.

18. Catherine O'Neill, "Saving the Spotted Owl," *The Washington Post*, July 3, 1990.

19. Yardley, "Plan Issued to Save Northern Spotted Owl."

20. National Council for Science and the Environment, "Beyond Old Growth: Older Forests in a Changing World." A synthesis of findings from five regional workshops, 2008, Washington, DC, 23.

21. Ciddi Veeresham, "Natural Products Derived from Plants as a Source of Drugs," *Journal of Advanced Pharmaceutical Technology & Research*, 3, no. 4 (October–December 2012).

22. David Yarnold, "Politicians Are Offering a False Choice Between Nature and the Economy, *Audubon* (summer 2017).

23. WWF, "CITES," https://www.worldwildlife.org/pages/cites.

24. National Parks Conservation Association, "Wolves at Isle Royale," https://www.npca.org/advocacy/37-wolves-at-isle-royale.

CHAPTER 13

1. Dan Egan, The Death and Life of the Great Lakes (New York: W. W. Norton and Company, 2017), xii.

2. The National Wildlife Federation, "The Great Lakes," https://www.nwf.org/Educational-Resources/Wildlife-Guide/Wild-Places/Great-Lakes.

3. Damian Carrington, "What Is Biodiversity and Why Does It Matter to Us?" *The Guardian*, March 12, 2018, https://www.theguardian.com/news/2018/mar/12/what-is-biodiversity-and-why-does-it-matter-to-us.

4. Edward O. Wilson, "The Biological Diversity Crisis," *BioScience* 35, no. 11 (1985): 701, http://www.jstor.org/stable/1310051.

5. Wilson, "The Biological Diversity Crisis," 700.

6. Camilo Mora, Derek P. Tittensor, Sina Adl, et al., "How Many Species Are There on Earth and in the Ocean?" *PLOS Biology*, August 23, 2011, https://journals.plos.org/plosbiology/article?id=10.1371/journal.pbio.1001127.

7. Mora, Tittensor, Adl, et al., "How Many Species Are There?"

8. Mora, Tittensor, Adl, et al., "How Many Species Are There?"

9. Food and Agriculture Organization of the United Nations, "The State of the World's Biodiversity for Food and Agriculture," 2019, https://www.fao.org/state-of-biodiversity-for-food-agriculture/en/.

10. Matt Hambly, "Biodiversity," New Scientist, https://www.newscientist.com/definition/biodiversity/#ixzz7BgKUzEPB.

11. UN Environment Programme, "The State of the World's Forests: Forests, Biodiversity

and People," May 22, 2020, https://www.unep.org/resources/state-worlds-forests-forests-biodiversity-and-people.

12. Katie Pavid, "What Is Biodiversity?" Natural History Museum, May 22, 2020, https://www.nhm.ac.uk/discover/what-is-biodiversity.html.

13. Hambly, "What Is Biodiversity?"

14. Elizabeth Kolbert, *The Sixth Extinction: An Unnatural History* (New York: Henry Holt and Company, 2014), Prologue.

15. Merriam-Webster, definition of "Faustian," https://www.merriam-webster.com/dictionary/Faustian.

16. Mark Vellend, "The Biodiversity Conservation Paradox," *American Scientist*, https://www.americanscientist.org/article/the-biodiversity-conservation-paradox.

17. Fon Mathuros, "New Partnership Aims to Sequence Genomes of All Life on Earth, Unlock Nature's Value, Tackle Bio-Piracy and Habitat Loss," World Economic Forum, January 23, 2018, https://www.weforum.org/press/2018/01/new-partnership-aims-to-sequence-genomes-of-all-life-on-earth-unlock-nature-s-value-tackle-bio-piracy-and-habitat-loss/.

18. UNESCO, "Ocean Life: The Marine Age of Discovery," https://en.unesco.org/news/ocean-life-marine-age-discovery-0.

19. Merriam-Webster, definition of "Microbiome," https://www.merriam-webster.com/dictionary/microbiome, attributed to Carl Zimmer.

20. *National Geographic*, "BioBlitz and iNaturalist: Counting Species through Citizen Science," https://www.nationalgeographic.org/projects/bioblitz/.

21. GBIF, Global Biodiversity Information Facility, Free and Open Access to Biodiversity Data, https://www.gbif.org.

22. Rebecca Gosling, "What Is a Seed Bank, How Does It Work and Why Is It Important?" Woodland Trust, December 2, 2020, https://www.woodlandtrust.org.uk/blog/2020/12/what-is-a-seed-bank.

23. Bee City USA, "No Mow May," https://beecityusa.org/no-mow-may.

24. Carrington, "What Is Biodiversity?"

25. Carrington, What Is Biodiversity?"

For Further Reading

Critical and Creative Thinking

Apps, Jerry. *Telling Your Story*. Wheat Ridge, CO: Fulcrum Publishing, 2016. A guidebook for personal storytelling, including suggestions for keeping a personal journal.

Brookfield, Stephen D. *Teaching for Critical Thinking*. San Francisco: Jossey-Bass, 2012. What is critical thinking? How critical thinking is learned; reading and writing critically.

Cohen, Martin, *Critical Thinking Skills for Dummies*. Hoboken, NJ: John Wiley, 2015. Focuses on developing critical thinking skills including how people think, applying critical thinking in practice, and presenting evidence and justifying opinions.

Dimmet, Ernest. *The Art of Thinking*. New York: Simon and Schuster, 1962. Possibility of an art of thinking. Nonthinking lives. Comprehending and critical reading. Writing as an aid to thought.

Haber, Jonathan. *Critical Thinking*. Cambridge, MA: MIT Press, 2020. Defining critical thinking, history of critical thinking, teaching and assessing critical thinking.

Jacobs, Alan. *How to Think*. London: Profile Books, 2018. P. 14: "Thinking is not the decision, but what goes into the decision"; p. 15: dealing with biases; p. 17: "Relatively few people want to think. Thinking troubles us; thinking tires us. Thinking can complicate our lives; thinking can set us at odds, or at least complicates our relationships, with those we admire or love or follow."

Kahneman, Daniel. *Thinking Fast and Slow*. New York: Farrar, Straus and Giroux, 2011. All about judgment and decision-making.

Paul, Richard, and Linda Elder. *The Miniature Guide to Critical Thinking*. Lanham, MD: Rowman and Littlefield Publishing Group, 2020. Why critical thinking? Stages of critical thinking development. Questions using elements of thought. A checklist for reasoning. Purpose; settle some question; solve some problem; all reasoning based on assumptions, data, information, and evidence.

Ruggiero, Vincent Ryan. *The Art of Thinking: A Guide to Critical and Creative Thought*. New York: Pearson Longman, 2007. The importance of thinking. A look at both critical and creative thinking and how to do each.

Von Oech, Roger. *A Whack on the Side of the Head: How You Can Be More Creative*. New York: Business Plus (Hachette Publishing), 2008. All about creative thinking and how to do it.

Environmental Concerns

Apps Jerry. *The Great Sand Fracas of Ames County*. Madison: University of Wisconsin Press, 2014. As one reviewer wrote: "Apps has tapped into a highly controversial issue to explore contemporary Midwestern values—historical preservation versus forces of change, environmental protection versus economic opportunity." The story is about a sand mining company attempting to locate in a small rural community.

Apps, Jerry. *Farm Winter with Jerry Apps*. PBS Wisconsin, 2013. An hour-long documentary. Winter in the upper Midwest. How winter affects us, including how we think and how we see the world.

Apps, Jerry. *Never Curse the Rain: A Farm Boy's Reflections on Water*. Madison: Wisconsin Historical Society Press, 2017. The author's reflections on the meaning of water in his life.

Apps, Jerry. *Settlers Valley*. Madison: University of Wisconsin Press, 2021. Apps shows how small acreage farming can heal those with emotional and physical challenges, and at the same time heal the land.

Apps, Jerry. *The Civilian Conservation Corps in Wisconsin: Nature's Army at Work*. Madison: Wisconsin Historical Society Press, 2019. The story of the CCC, which operated from 1933 to 1942, when out-of-work young men improved soil conservation, restored state and national parks, and did much to restore the forests that had earlier been decimated during the logging era.

Apps, Jerry. *The Land Still Lives*. Madison: Wisconsin Historical Society Press, 1970, 2019. With a foreword by Gaylord Nelson, founder of Earth Day, Apps tells the story of a farm he and his family purchased in 1966.

Apps, Jerry. *The Quiet Season*. Madison: Wisconsin Historical Society Press, 2013. All about winter in the upper Midwest from a personal perspective.

Apps, Jerry. *The Travels of Increase Joseph*. Madison: University of Wisconsin Press, 2003, 2010. A novel whose main character, Increase Joseph Link, a preacher, calls on his people to protect the land.

Apps, Jerry. *Whispers and Shadows: A Naturalist's Memoir*. Madison: Wisconsin Historical Society Press, 2015. Apps learned from his father to listen to the whispers and look in the shadows when out in nature—seek out those things not easily seen, and listen for the subtle sounds often not heard.

Apps, Jerry. *When the White Pine Was King: A History of Lumberjacks, Log Drives, and Sawdust Cities in Wisconsin*. Madison: Wisconsin Historical Society Press, 2020. A history of the logging industry from the 1800s to the present. An examination of the cutover after the loggers left and the environmental damage that was caused.

Apps, Jerry. *The Land with Jerry Apps*. PBS Wisconsin, 2015. An hour-long documentary. In his first two PBS specials, historian Apps explored his childhood and country winters—from the Great Depression to World War II in Waushara County, through personal memories and photos from the Wild Rose community.

Boylan, Michael, ed. *Environmental Ethics*. Hoboken, NJ: Prentice Hall, 2001. An anthology that contains material on sustainability and climate change.

Carson, Rachel, *Silent Spring*. New York: Houghton Mifflin, 1962, 1994. Carson carefully researched the side effects of DDT. The book spurred a revolution in thinking about the need for laws to protect air, land, and water.

Davis Hanson, Victor. *Fields Without Dreams: Defending the Agrarian Idea*. New York: The Free Press, 1996. Intimacy between people and the land. "As the family farm all but vanishes in our nation, it is neither food production nor the environment that will most suffer—but rather our nation will lose its last real connection with the virtues and work ethic . . . upon which American Society rests."

Egan, Dan. *The Death and Life of the Great Lakes*. New York: W. W. Norton and Company, 2017. The book provides a portrait of an ecological catastrophe happening right before our eyes, blending the epic story of the lakes with an examination of the perils they face and the ways we can restore and preserve them for generations to come.

Fox, Stephen. *The American Conservation Movement: John Muir and His Legacy.* Madison: University of Wisconsin Press, 1981. Focusing on the work of famed naturalist, John Muir, Fox examines the development of the conservation movement from 1890 to 1975.

Freyfogle, Eric T. *The New Agrarianism: Land, Culture, and the Community of Life.* Washington, DC: Island Press, 2001. "Marked resurgence of agrarian practices and values in rural areas, suburbs, and even cities." Strengthening our ties to the land. "Reaching beyond food production to include a wide constellation of ideals, loyalties, sentiments and hopes."

The Future of Farming and Rural Life in Wisconsin. Madison: Wisconsin Academy of Sciences, Arts and Letters, 2007. Status of rural Wisconsin, sustaining our communities; Food systems: the Wisconsin Advantage, (the future farm, p. 140); the land we tend; production agriculture: past, present, future.

Gates, Bill. *How to Avoid a Climate Disaster: The Solutions We Have and the Breakthroughs We Need.* New York: Alfred A. Knopf, 2021. Gates sets out a plan to stop the planet's "slide to certain environmental disaster."

Gore, Al. *Earth in the Balance: Ecology and the Human Spirit.* New York: Penguin, 1992, 1993. Gore argues that the "engines of human civilization" have brought us to the brink of catastrophe.

Groh, Trauger, and Steven McFadden. *Farms of Tomorrow Revisited: Community Supported Farms—Farm Supported Communities.* Kimberton, PA: Biodynamic Farming and Gardening Association, 1997. A guide for community supported agriculture (CSA). Includes practical examples for those wishing to become a part of the movement.

Gundersen, Adolf G. *The Environmental Promise of Democratic Deliberation.* Madison: University of Wisconsin Press, 1995. Gundersen argues that ordinary citizens can do much more in helping solve many environmental challenges.

Hart, John Fraser. *The Land That Feeds Us.* New York: W. W. Norton, 1991. A look at farming's past, and what farming will look like in the next century.

Higbee, Edward. *Farms and Farmers in an Urban Age.* New York: The Twentieth

Century Fund, 1963. Contents: The technological revolution, who is the American farmer? The farm in the American mind, land, and the demand for space.

Hovel, Joe. *Northwoods Forest Conservation: Managing Forestlands for the Future.* Conover, WI: Partners in Forestry Cooperative and Northwoods Alliance, 2021. A practical look at the challenges forests face and how to combat them.

Kimbrell, Andrew, ed. *The Fatal Harvest Reader: The Tragedy of Industrial Agriculture.* Washington, DC: Island Press, 2002. A collection of essays that "depict the current industrial agricultural system and its devastating impacts on the environment, human health, and farm communities."

Kolbert, Elizabeth. *The Sixth Extinction: An Unnatural History.* New York: Henry Holt and Company, 2014. The book argues that the planet is in the midst of a modern, human-made, sixth extinction. In the book, Kolbert chronicles previous mass extinction events and compares them to the accelerated, widespread extinctions during our present time.

Kolbert, Elizabeth. *Under a White Sky: The Nature of the Future.* New York: Penguin Random House, 2021. Looks at humanity's transformative impact on the environment and asks: After doing so much damage, can we change nature, this time to save it?

Krimsky, Sheldon, and Jeremy Gruber. *The GMO Deception: What You Need to Know about the Food, Corporations, and Government Agencies Putting Our Families and Environment at Risk.* New York: Skyhorse Publishing, 2014. A comprehensive look at genetically modified organisms, defined as any organism whose genetic material has been altered using genetic engineering techniques. The authors offer an in-depth look at the social, political, and ethical implications of genetically modified food.

Leopold, Aldo. *A Sand County Almanac.* New York: Oxford University Press, 1949. A classic in the environmental world. On these pages, Leopold introduced the concept of a land ethic.

Leopold, Aldo. *For the Health of the Land.* Washington, DC: Island Press, 1999. In this collection of essays, Leopold develops the concept of "land health." The essays are an action-based plan to restore land to ecological health.

Logsdon, Gene. *At Nature's Pace: Farming and the American Dream.* New York: Pantheon Books, 1994. Today's single-crop megafarms, and the urban communities that depend on them, are headed toward an economic and biological crisis. "Healthy agricultural practices can work only at nature's pace, grounded in reverence for the land, biological efficiency that transcends technological shortsightedness."

McKibben, Bill. *The End of Nature*. New York: Random House, 1989, 2006. McGibbon argues that the survival of the planet depends on a philosophical shift in the way we relate to nature.

Montgomery, David R. *Growing a Revolution: Bringing Our Soil Back to Life*. New York: W. W. Norton and Company, 2017. Montgomery, "cutting through standard debates about conventional and organic farming . . . explores why practices based on principles of conservation agriculture help restore soil health and fertility."

Muir, John. *The Story of My Boyhood and Youth*. Boston: Houghton Mifflin, 1913.

Nash, Roderick F. *The Rights of Nature: A History of Environmental Ethics*. Madison: University of Wisconsin Press, 1989. Contains the history of the origins of environmental ethical thought with a helpful study of the impact on conservation thinking of pioneers such as John Muir, Gifford Pinchot, and Aldo Leopold.

Nelson, Gaylord. *Beyond Earth Day: Fulfilling the Promise*. Madison: University of Wisconsin Press, 2002, 2012. The founder of Earth Day discusses the modern-day environmental movement, detailing the planet's most critical concerns.

Nelson, Rebecca L. *Aquaponic Food Production: Raising Fish and Plants for Food and Profit*. Montello, WI: Nelson and Pade, Inc., 2008. Everything from the history of aquaponics, to detailed instructions on how to do it. Nelson discusses fish species as well as plants that do well in aquaponic growing situations.

Pollan, Michael. *The Omnivore's Dilemma*. New York: Penguin, 2006. How what we eat affects the environment.

Pyle, George. *Raising Less Corn, More Hell: The Case for the Independent Farm and Against Industrial Food*. New York: Public Affairs, 2005. America's "breadbasket" is increasingly controlled by large corporations. They push farmers further into debt. They pollute the Earth and wear out the soil. Pyle concludes that unless we do something about this dilemma, we will turn much of rural America into a "permanent environmental and economic wasteland."

Rutstrum, Calvin, *The Wilderness Life*. New York: Macmillan, 1975. Explores the importance of looking at our past as we consider the future.

Schmidtz, David, and Elizabeth Willott, eds. *Environmental Ethics: What Really Matters, What Really Works*. New York: Oxford University Press, 2002. In addition to classic and contemporary readings in environmental ethics, this volume includes coverage of biological, social-scientific, management, and activist approaches to environmental problems.

Shepard, Mark. *Restoration Agriculture*. Greeley, CO: Acres U.S.A., 2013. Our present reality, challenges facing agriculture, farming in nature's image, steps toward restorative agriculture, transitional strategy, and so on. "Reveals how to sustainably grow perennial food crops that can feed us in our resourced-compromised future."

Suzuki, David, The *Sacred Balance: Rediscovering Our Place in Nature*. Toronto: Greystone Books, 1997, 2002. Suzuki makes the point that we are all a part of nature, not apart from it.

WASAL. *Waters of Wisconsin: The Future of Our Aquatic Ecosystems and Resources*. Madison: Wisconsin Academy of Sciences, Arts and Letters, 2003. A report based on a three-year initiative to review the water situation in Wisconsin.

Wirzba, Norman, ed. *The Essential Agrarian Reader: The Future of Culture, Community, and the Land*. Lexington: University Press of Kentucky, 2003. Contributors to this book suggest how our society can take "practical steps toward integrating soils, watersheds, forests, wildlife, urban areas, and human populations into one great system."

Index

About the Authors

Jerry Apps is a former county extension agent and is now professor emeritus at the University of Wisconsin–Madison, where he taught for thirty years. Today he works as a rural historian and full-time writer and is the author of many books on rural history, country life, and the environment. He has created six hour-long documentaries with PBS Wisconsin, has won several awards for his writing, and

Photo by Steve Apps

won a regional Emmy Award for the TV program *A Farm Winter*. Jerry and his wife, Ruth, have three children, seven grandchildren, and three great-grandsons. They divide their time between their home in Madison, Wisconsin, and their farm, Roshara, in Waushara County.

Natasha Kassulke is a former journalist for the *Wisconsin State Journal* and former editor of *Wisconsin Natural Resources* magazine. Today, she directs communications for the Office of the Vice Chancellor for Research and Graduate Education at the University of Wisconsin–Madison and teaches journalism courses part-time at Madison College. She and her husband, Steve Apps, live in Madison, Wisconsin.

Photo by Steve Apps